Tibaldo

and the

Hole

in the

Calendar

Abner Shimony
Illustrated by Jonathan Shimony

Tibaldo and the Hole in the Calendar

COPERNICUS
An Imprint of Springer-Verlag

Published in the United States by Copernicus, an imprint of Springer-Verlag New York, Inc.

Copernicus
Springer-Verlag New York, Inc.
175 Fifth Avenue
New York, NY 10010
USA

Library of Congress Cataloging-in-Publication Data
Shimony, Abner.
Tibaldo and the hole in the calendar / Abner Shimony : with illustrations by Jonathan Shimony.
p. cm.
ISBN 0-387-94935-6 (alk. paper)
1. Bologna (Italy)—History—Papal rule, 1506-1797—Fiction. 2. Astronomy—Italy—History—16th century—Fiction. 3. Gregory XIII, Pope, 1502-1585—Fiction. 4. Calendar, Gregorian—History—Fiction. I. Title.
PS3569.H492T5 1997 97-2157
813'.54—dc21

Manufactured in the United States of America.
Printed on acid-free paper.

9 8 7 6 5 4 3 2 1

ISBN 0-387-94935-6 SPIN 10557994

to the memory of
Annemarie Anrod Shimony
and Miriam Gail Farber

Acknowledgments

We are endebted to many people for their expertise and generous assistance. Ethan Shimony lent his face and temperament to our composition of Tibaldo. Howard Stein greatly improved the text by his astronomical and historical corrections and his meticulous and sensitive emendations of language. Ruth Montgomery enriched the story by suggesting the addition of feminine characters, particularly Tibaldo's oldest sister. Teresa Shields, William Frucht, and Jeremiah Lyons of Springer-Verlag properly insisted upon amplifying both the story and the scientific material and cooperated in achieving the amplification. Father Reginald Foster translated an English version of Pope Gregory's codicil to the reform of the calendar into authentic papal Latin, and Katherine Geffcken wrote to Rome asking him to do the translation. George Greenstein corrected some astronomical errors, and Robert Palter made suggestions regarding Renaissance medicine. Don Howard corrected some errors in chronology, and John Ongley tracked down the complete text of Pope Gregory XIII's proclamation of the new calendar. Ernesto Corinaldesi and David Goldberger made useful suggestions concerning Italian proper names. The American College of Nurse-Midwives supplied valuable literature on midwifery. Karen Phillips was a technically accomplished director of design and a very able collaborator in preparing the layout of the book. Alain Cazalis and Michel Viot provided helpful artistic advice and access to the printing facilities at the Duperré School of Applied Arts in Paris. Ruth Rogers was a helpful guide to the Renaissance books in the Special Collections of the Wellesley College Library. Odile Simon was a helpful guide at Duperré to Renaissance images and literature. Nigel Freeman provided a crucial workplace in Brooklyn for the final stages of illustration. Emma Brante supplied technical and moral support throughout the art work. David Gould steadily encouraged the making of illustrations. Signori Sanzio, Vecellio, and Caliari provided extraordinary instruction and inspiration in Renaissance drawing.

ABNER SHIMONY
JONATHAN SHIMONY

Contents

A Note on Fact and Fiction

Tibaldo Bondi and all of his relatives are fictional, as are the school of St.-Joseph-in-the-Corner and its principal, teachers, students, and their families. Governor Domitiani and his officers, Signora Guardabassi, and il Torrentino are fictional. All other characters mentioned by name are historical persons, but some of their adventures in the story are fictional. Pope Gregory, for example, did not travel to Bologna near the end of his life, nor did he modify his proclamation of the new calendar. The real Professor Turisanus lived in the fourteenth not the sixteenth century, but his character has not been much changed by this transposition in time. In the exposition of science and history our aim was to achieve accuracy without technical difficulty. The appendix, "More and Better Astronomy," is not essential to the story of Tibaldo, as are the astronomy and history in the main text, but we hope that it will be read for its scientific interest.

The World of Tibaldo

There are three kinds of worlds: the great world, the middle-sized world, and many little worlds. The great world is the world of nature, consisting of the stars, the sun, the planets, the moon, the earth, and all that the earth contains. The middle-sized world is the world of human society, with its nations, governments, armies, religions, factories, farms, schools, families and everything else formed by human beings. The many little worlds are individual human beings. Each man, woman and child is a little world, but of course each is shaped and influenced by the middle-sized world of human society, often in strange and surprising ways. Likewise, each is shaped and influenced by the great world of nature, sometimes in ways that no one could predict.

This story is about a particular little world—a boy named Tibaldo Bondi, who was born in Bologna in the north of Italy on October 10, 1570. The exact date of his birth is very important for our story, since it was responsible for some strange things that happened to him. Had he been born on October 1, 1570, or October 20, 1570, his adventures would have been very different. The strangest of Tibaldo's adventures occurred in 1582, which is the year in which Pope Gregory XIII declared that the calendar had to be changed. In typical adventure stories soldiers fight enemies, or knights slay dragons, or explorers overcome hardships, or astronauts face dangers in space

This adventure story, however, is different.
In it a small boy fights a battle against a calendar.
Why does he do that? How can any one do it?
Many things have to be explained in order
to show how Tibaldo could take up
a battle against a calendar.

The family into which Tibaldo was born was large and active. His parents, Lorenzo and Teresa Bondi, had eight children, three girls and five boys. Tibaldo was their youngest child. Since the parents were poor, their house was small and therefore crowded and busy, even after Tibaldo was sent away to school and came home only on Sundays and holidays. And when the girls married and moved to homes of their own, it still was crowded, especially after the oldest daughter, Anna Maria, became a widow and had to move back into her parents' house with her two small sons. Tibaldo was particularly fond of Anna Maria and her sons, and he was proud to be an uncle,

greatly admired by his nephews, though he was only slightly older than they.

Tibaldo's mother, Teresa, had a great talent for calming the uproar of her household and making peace among the many squabbling children. Furthermore, she made sure that no one was neglected. Each child was made to feel important, useful, and necessary to the daily life of the Bondi household. One of her strategies was to celebrate the birthday of each child and grand-child with a lavish feast. You might think that her strategy was not unusual, but as a matter of fact birthdays were not much celebrated in Italy at that time. Tibaldo's mother insisted upon a celebration because she felt that every one should have the experience of being the most important person in the family for one day of the year. The pride of that day, she said, would give the child self-confidence for the entire year. Consequently, she prepared an elaborate dinner with pasta, sausages, salads, cakes, and fruits. The birthday child was propped up with cushions on the largest chair of the house to make it look like a throne, and she made a brightly colored cloth and cardboard crown to place upon the child's head. The child was king or queen for the day.

Lorenzo Bondi worked at the Medical School of the University of Bologna, which was the most famous medical school in all of Europe and one of the oldest. He was therefore very proud to work there, and his son Tibaldo was even prouder. But it must be explained that Lorenzo Bondi was not a professor at the University and in fact was not a doctor at all. How could he be, when he did not understand Latin and could not even read Italian well? He was not able to explain diseases as the professors did in their lectures to the medical students. They would say, for example, that a person becomes sick when there is not a proper balance of hot and cold, moist and dry in his body. Or they would assert that it is the influence of the planets that causes the flu (the word flu is just a short form of the Italian word *influenza,* which means influence). Or they would explain that a person becomes mentally sick because his mind is disturbed by the moon (and that is why a mentally sick person is sometimes called a lunatic, for the Latin word for the moon is *luna*). No, Lorenzo Bondi was a simple and modest man. He was nothing more than the assistant to the great, learned, world-famous Professor Petrus Turisanus, and the Professor did indeed give lectures that were full of long words and grand theory.

What did Lorenzo Bondi do as the assistant to Professor Turisanus? Only simple things, and only things with his hands. If a soldier was carried into the hospital with a deep cut from a sword, Professor Turisanus would be called upon to make a diagnosis. He would say that it is a great danger to the balance of elements in the body if much blood flows from the wound, so that moisture and heat are lost, and an even greater danger if an infection sets in causing too much heat. After he finished his diagnosis, which was the most important part of the treatment, he would leave the soldier in the hands of Lorenzo Bondi, who would take care of details. Lorenzo would stop the flow of blood by pressing on an artery above the wound. When the blood stopped flowing, he would carefully wash the wound with wine, which was the best substance known at that time for preventing infection.

Then he would put on clean bandages to close and protect the wound, and he would change them three times a day. Lorenzo had a gentle way of doing these things, so the pain was somehow less than when the assistants of other doctors did similar things. When someone came to Professor Turisanus with a broken arm, the Professor hardly looked at it, because the problem was merely mechanical and hence not worthy of his knowledge. It was Lorenzo who carefully and quickly set the bone and prepared a splint to hold the arm straight until the break healed. Almost always the arm became strong and straight again. Therefore Professor Turisanus became famous for his treatments of wounds and of broken bones, and patients came to him not only from Bologna but from far-off places like Florence and Pisa. A good reputation spreads quickly.

From the time Tibaldo was five years old he was very interested in his father's work and used to beg to be taken to the Medical School. That would not be allowed today, because there are now strict rules about keeping curious people, especially children, out of the way when a sick person is being treated. But the rules were not as strict in those days, and so Lorenzo was able to let his son come to watch him work with patients. Tibaldo thought his father a great man, even though Lorenzo— who was modest and not well educated— regarded himself as a simple assistant. But Tibaldo kept his eyes open, saw many things, and thought for himself, and he was right in his opinion of Lorenzo. Tibaldo dreamt that some day he too could cure people as his father did.

When Tibaldo was seven years old an important thing happened to him on one of his visits to the Medical School. He saw a large group of students in long gowns trooping into a hall that had a high ceiling, long benches, and a high lecture stand at the front. At first there was no one at the lecture stand, but then a tall, white-bearded man wearing a magnificent black and gold coat and a velvet hat walked with great dignity to the lecture stand. It was Professor Petrus Turisanus himself. All the students respectfully rose for him until he told them to sit down for the lecture. In a strong, solemn, slow voice the Professor lectured for an hour about many things, using words that Tibaldo had never heard before, like *element* and *compound, affinity* and *discord, generation* and *corruption, opposition* and *conjunction of the planets,* the *microcosm* of the human body and the *macrocosm* of the heavens. He quoted whole pages from famous Greek and Latin writers like Aristotle, Hippocrates, and Galen. Tibaldo understood very little, but he was fascinated. He listened so attentively that he could remember almost everything the

Professor had said in Italian, and he greatly regretted missing the Latin and Greek.

When the lecture ended the students again rose respectfully, and Professor Turisanus walked with great dignity to the door, not looking to the right or left. His entrance and exit from the lecture were ceremonies which he greatly enjoyed. Out of the corner of his eye, he saw Tibaldo. He was surprised to see a little boy, half the size of his students, quietly standing in the back of the lecture hall. He did not recognize Tibaldo as the son of his assistant, and indeed he didn't not even know that his assistant had any children. Professor Turisanus stopped and called the boy to him. Of course Tibaldo was frightened, because he knew that he was not supposed to be in the lecture hall, but he was courageous enough not to run away.

Professor Turisanus asked Tibaldo, "Boy, did you hear my lecture?" Tibaldo said yes, and then the Professor said, "My lectures are not meant to be wasted. You will be punished if you wasted one. Let me test you. What are the fluids of the body, according to Hippocrates? Quick, answer." Tibaldo could not have been more startled. He was so frightened by the world-famous man towering above him and glaring that he could hardly have told his own name. Still, he had listened to the Professor's lecture with deep attention and began to answer as if talking in his sleep: "The body of man has in itself blood, phlegm, yellow bile, and black bile; these make up the nature of his body, and through these he feels pain or enjoys health. Now he enjoys the most perfect health when these elements are duly proportional to one another in respect to compounding, power, and bulk, and when they are perfectly mingled Pain is felt when one of these elements is in defect or excess." Actually, Tibaldo could have continued like this for many more minutes, repeating what he had heard with very little understanding, like a record. But Professor Turisanus was astonished and stopped him, saying, "How old are you, boy?" Tibaldo did not answer at once, because he had not just memorized the answer to that question, but finally he managed to say, "Seven years old, sir." Then Professor Turisanus said, "I do not know who you are, but I know what you are. You have a mind for medicine, and your mind must be cultivated. And I shall make sure that you receive the education you deserve."

There was a thought in the Professor's mind that he would not have admitted to any one. He had seven daughters and no sons, and in those days a woman was not allowed to become a doctor and was rarely allowed to receive any kind of advanced education. This state of affairs was painful to Professor Turisanus, because he wanted to pass on all his immense knowledge to a son, who would continue his name and fame. It was even more painful because thirty years earlier he had delivered a famous series of lectures at the University of Bologna and the University of Padua on how to control things in such a way as to have sons rather than daughters. In recent years, however, he had stopped giving that series of lectures. The secret thought came into the mind of Professor Turisanus that Tibaldo would become the son he had never had. The fact that Tibaldo was the child of some one else—who turned out to be his own assistant, Lorenzo Bondi—made no difference to his secret thought.

Tibaldo's School

Professor Turisanus kept his word. He arranged that Tibaldo be admitted to the best school in Bologna, the School of St.-Joseph-in-the-Corner, and he paid all of the expenses of tuition, room, and board. The principal of the school was Master Domenico, who was famous for educating boys (no girls) so well that many were thoroughly equipped to enter the University of Bologna. The boys who entered St-Joseph-in-the-Corner knew little Latin, but by the time they graduated they knew Latin well. They also learned some Greek, a little arithmetic, and were introduced to astronomy.

We cannot say that Tibaldo was happy at his school. Every day except Sundays and holidays he had to live away from his home and family, both of which he loved dearly. The school was a place of discipline. Master Domenico boasted, "There is no room here for chattering blue-jays, or for slow turtles, or for jumping rabbits, or for feeble chickens, or for stupid donkeys, or for trick-playing monkeys." He was far from gentle and believed that an occasional birch whipping made his students con-

centrate on their Latin, a belief that some of the other teachers shared. Tibaldo found his studies less interesting than watching his father set broken bones and take care of wounds. Still we cannot say that he was unhappy. His parents were very proud that Tibaldo had attracted the attention of the great Professor Petrus Turisanus, and they were pleased that he would learn all the mysterious things that one needed to know in order to be a medical doctor. Tibaldo was proud of the future that was being prepared for him. Finally, he had good friends at school, because he was a lively boy.

During the first hour of classes the boys would have to recite whole pages which they were supposed to have memorized the previous evening. Sometimes there were pages of rules of Latin grammar. Sometimes there were pages from the Bible in Latin. Often there were speeches by the great Roman orator Cicero. Occasionally the boys had to recite from a book of Roman law (for Master Domenico liked to say that many of his students had become lawyers), and sometimes from Galen's book on diseases (because Master Domenico liked to tell how many of his students had become medical doctors), and frequently they recited from a book of logic (for he was proud that he had produced well known scholars). When a boy made too many mistakes in his recitation Master Domenico used his birch switch. Whether that made the students' minds stronger may be doubted, but perhaps it frightened them into working harder.

Another hour was spent on translation. One of the assistant teachers would read rapidly in Latin, and the boys

would have to write on their slates the translation in Italian, their usual language. The teacher would then read to them in Italian, and they would have to translate as rapidly as possible into Latin.

The next hour was spent doing arithmetic, but always using Roman numerals, in which I is 1, V is 5, X is 10, L is 50, C is 100, D is 500, M is 1000, and a bar is placed over a letter to indicate multiplication by a thousand. The number 488, for example, is written in

Roman numerals as

CCCCLXXXVIII,

and the number 877 is

DCCCLXXVII.

As you know, it is not very difficult to multiply 488 by 877 to find 427,976, in the way we do it today. But try multiplying

CCCCLXXXVIII by DCCCLXXVII

to find the result

$\overline{\text{CCCCXXVII}}$DCCCCLXXVI,

and you will have a taste of how difficult arithmetic was in those days.

One hour was spent on debating, which Master Domenico supervised himself. He used to say that nothing sharpens the mind of a young man more than making an argument about a subject he knows nothing about and finding weak spots in an opponent's argument. He would choose two boys, place them on a platform in front of the class, and then pose an impossibly hard question. For example, "Are there both male and female angels?," or "If a widow marries a second husband and later goes to heaven, does she live with her first or her second husband?" or "Is it easier to turn gold into lead than lead into gold?" One boy had to defend one point of view and the other had to defend its opposite. Of course, the debate had to be in Latin, and good grammar was as important as sharp reasoning to win the debate and escape the birch.

Master Domenico used to say that his school taught athletics for the mind. Reciting long passages by heart was like weight-lifting. Translating quickly from Latin into Italian and from Italian into Latin was like running races forward and backward. And debating was like wrestling. Although the boys were turned into good mental athletes, they were given little time for athletics of the body. There were no play periods. Sometimes the students could slip away for a few minutes before or after meals or on the way to church to release some of the energy stored up in their arms and legs. Altogether, it was not an easy experience for a young boy.

Tibaldo did very well in his studies, but it cannot be said he enjoyed all of them. They had to be done, because his chance of becoming a medical doctor depended upon doing well. He knew also that his parents and Professor Turisanus were relying on him. Nevertheless, the memorizing and translating were drudgery to him. Occasionally he enjoyed debating, because he was quick-witted and often could win by thinking of an amusing argument.

One subject that did give Tibaldo pleasure was Greek. The reason was not that he liked learning yet another language but that the Greek teacher, Master

Demetrios, was a remarkable man. His great-grandfather had come to Italy from Constantinople in 1453. That was an important year in history, for it was then that the Turks captured Constantinople, which had been the capital of the Byzantine Empire, also known as the Roman Empire of the East. When the Turks pushed westward from central Asia they conquered more and more pieces of the Byzantine Empire until almost nothing remained but the capital city itself. Constantinople fought its last battle bravely, but finally it too was captured. Like many soldiers from Constantinople, Demetrios's great-grandfather fled to Italy rather than live in the Turkish Empire. He had a choice between serving as a professional soldier in the army of one of the many small Italian cities and duchies or teaching Greek, which had been the language of the Byzantine Empire. Greek was in much demand because many of the great books of antiquity had not been translated into Italian or Latin; furthermore originals were better than

translations. The former soldier became a Greek teacher because he did not want to go to war except in a good cause. His son and grandson had followed in his new profession, and now his great-grandson Demetrios was also a teacher of Greek. The memory of the past was kept alive in Demetrios's family. He knew about all the glorious battles in which his great-grandfather had fought, and he liked to tell his students about them. The students soon learned that if they were translating some passage from Homer about the Trojan War and were tired of working out the grammar, they could distract their teacher by asking military questions. For example, what is the best way to prevent enemy soldiers from climbing over the walls of a fortress? When asked this question Demetrios would start to lecture on archery and burning sulfur, and there was no more Greek grammar during that class period.

In fairness to the students we should say that they were not just trying to get out of work—though that was one motivation. They were also fascinated by Demetrios's enthusiasm and his vivid descriptions. He had not forgotten Constantinople even though he had never visited there, and he was sure that the time would come for a great army to rescue the city from the Turks and restore the Byzantine Empire. If that did not happen in his own lifetime, then surely it would in the lifetime of his children or his grandchildren. The students in Demetrios's class were inspired by his determination, and they too wanted to do great deeds. He told them that they should never be afraid of anything. Tibaldo was more inspired to be courageous than any one else in the class.

There was another class that Tibaldo enjoyed because of the subject itself, and that was astronomy. The reason astronomy was taught at the school of St.-Joseph-in-the-Corner is very interesting. Master Domenico, the principal, had little curiosity for the great world of nature. What he cared about was the reputation of his school for preparing boys to enter professions in which they would become famous and wealthy. He knew that many professional men believed in astrology, which taught that the planets and the constellations influence the health and success of individual human beings. According to the astrologers there are spirits associated with the sun, the moon, each planet, and each star. At the moment of birth of a baby, each of these heavenly bodies has a definite

location relative to the place of birth, and the spirits asscociated with them influence and even control the life of the newborn baby. During the subsequent life of the baby many decisions will have to be made: whom to marry, where to reside, what occupation to choose, what exact times to launch important enterprises like a voyage or an investment or a battle. Which decisions would be fortunate and which would be unfortunate depend on the various influences of the spirits attached to the heavenly bodies, and exactly what those influences are depend on the pattern of the heavenly bodies at the moment of birth. Consequently, according to astrology, one needs experts—namely astrologers—to give advice for making wise decisions. Medical doctors needed to be astrologers themselves or at least to have advice from astrologers. Likewise, princes, generals, merchants, lawyers, and all other people making important decisons needed to consult astrologers. We must point out, however, that there is no evidence whatever that spirits are attached to the heavenly bodies. Furthermore careful observers have kept track of several children born at the same time—and therefore with the same pattern of heavenly bodies supposedly controlling their lives—and have found that the subsequent lives of these children were entirely different. Eventually most intelligent people realized that astrology was nonsense. For example, only about thirty years after Tibaldo's birth Shakespeare had one of his characters in a play shrewdly say,

"The fault, dear Brutus, is not
in our stars,
But in ourselves, that we are
underlings."

But Shakespeare was ahead of his time and besides was extraordinarily intelligent. During the period of Tibaldo's schooling the evidence against astrology had not yet been completely gathered, and therefore many influential people believed strongly in astrology. Master Domenico was not a man who thought much for himself, but since he was mainly concerned with the reputation of his school, he simply accepted the opinion that astrology is correct and that his students would eventually need to know about it. But why then was astronomy rather than astrology taught at the school of St.-Joseph-in-the-Corner? The answer is that astronomy is a study (among other things) of the paths that the sun, the moon, the planets, and the stars appear to make through the heavens as the hours of the day and night pass, and as the year progresses. The astrologers depended on astronomy in order to provide information about these paths. They used the correct information provided by astronomers in order to concoct their own nonsensical predictions about the influence of the heavenly bodies upon individual human beings. The students at the school of St.-Joseph-in-the-Corner were luckier than they realized, because Master Domenico thought that astronomy is an elementary subject that schoolboys can understand, whereas astrology is a profound and difficult subject that has to be learned at the university. Consequently, the students had occasional astronomy lectures during the day and viewings of the moon, stars, and planets at night, and were spared the nonsense of astrology. Furthermore, their teacher, who

was privately very skeptical of astrology but thought it
wise to keep his opinions to himself, was spared
the pain of having to teach something
that he did not believe in.

Lessons in Astronomy

The teacher of astronomy at the school of St.-Joseph-in-the-Corner was Master Vittorio Rhaeticus. He had been born in Poland and educated there, particularly by his grandfather Georg Joachim Rhaeticus, who was a famous astronomer and the pupil of an even more famous astronomer, the great Nicolaus Copernicus. Master Vittorio's first name was really Wojciech, but since that name was too difficult for the people in Bologna to pronounce, he accepted—with a little annoyance—the substitute Vittorio. What was much more disturbing to him was that his connections to his famous grandfather and his grandfather's great teacher were not advantageous to him in Bologna, as he had expected, but just the opposite. Master Vittorio had come to Bologna with letters of recommendation praising his knowledge of astronomy and his skill at performing mathematical calculations. He hoped to be appointed a lecturer in astronomy at the University of Bologna and eventually to become a professor. After all, Copernicus himself had attended the University of Bologna between 1497 and 1500 and was one of the most renowned people ever to study there. The trouble was that Copernicus was too famous. He had written a great book, *On the Revolutions of the Heavenly Spheres,* in which he reasoned that the earth revolved in an orbit around the sun once a year and indeed was a planet, like Mercury, Venus, Mars, Jupiter, and Saturn. He also proposed that the earth rotates from west to east about an axis once a day, and this rotation is responsible for the daily appearance of the sun rising in the east and setting in the west. Copernicus was bold and revolutionary, since for fifteen hundred years astronomers had generally accepted the theory that the earth is at the center of the universe, and that the sun, planets, and stars all revolve about it Copernicus's ideas were rejected not only by most astronomers of his time but also by the Church. According to the Church, all statements of the Bible are true, including those which imply that the earth is

fixed and the sun moves around it. Consequently, when Vittorio Rhaeticus came to Bologna hoping to present the ideas of his grandfather and of Copernicus at the University of Bologna, he was turned down because he was too radical. The only teaching position he could find in Bologna was at the school of St.-Joseph-in-the-Corner, where he was not expected to discuss theories at all, but merely to teach the students how to look up at the heavens and describe what they saw.

Tibaldo's first astronomy lesson took place on a clear evening in mid-September. The class ascended a narrow winding staircase to the roof of the school, which was large and flat and high enough above the surrounding buildings to give an unobstructed view of the heavens in almost all directions. Master Vittorio let the students gaze for a while without saying anything, and then he pointed out what they had already noticed—how still the dome of the heavens is. Then he said, "You must learn to be at home in the heavens. You should know your way around by recognizing familiar markers, just as you know your way around the city of Bologna." The markers he pointed out to them were some of the constellations, which are groupings of stars forming distinctive patterns. He showed them the Big Dipper, also known as the Big Bear, consisting of three stars in its handle and four in the bowl. Then he showed them Cassiopeia, which looks like the letter *W*. Overhead there was a constellation in the form of a cross, but with some imagination this cross begins to look like a flying swan. The name of this constellationis Cygnus, which is the Latin word for swan. Near the western horizon, where the sun had set only an hour or so previously, he pointed out a constellation he wanted the class to pay particular attention to. It was called Scorpio, and indeed if one used imagination the pattern would begin to resemble a scorpion. Every one noticed a very bright red star in Scorpio, called Antares—one of the things that Master Vittorio wanted the class to notice but not the most important. He had the students look for a long time, and then some one exclaimed: "Antares and the rest of Scorpio are going down over the horizon. They are setting!" And so they were. The students' first impression that everything is still in the night sky turned out not to be correct. The pattern of the stars was not fixed but slowly moving, and the movement was most obvious in the stars closest to the west-

ern horizon, where they were setting, or the eastern horizon, where they were rising.

After several nights of observation it became clear that as the night progressed the positions of the stars relative to each other did not change. The stars seemed to be attached to a great spherical dome above the earth and as the dome turned all the stars turned with it, always keeping the same distances from each other. Master Vittorio explained that the moving dome was only an illusion, and it is the turning of the earth itself that makes the whole pattern of the stars appear to rotate.

Some of Master Vittorio's astronomy classes took place in the daytime, when the students could observe the sun. The brighter students realized that if the rotation of the earth causes the appearance of turning of the pattern of the stars at night, the same turning should be responsible for the apparent movement of the sun through the sky during the day. One student asked whether the position of the sun relative to the stars is fixed, just as the positions of each star relative to all the others is fixed. Master Vittorio said, "That is a good suggestion, but it is not exactly correct.

From one day to the next the position of the sun relative to each star changes very little, and hence the sun at first appears to be fixed relative to the pattern of the stars. But over the course of a year its position changes very much. As a matter of fact, over the period of a year the sun goes around a circular path through the pattern of the stars, a path called the ecliptic. There are twelve constellations located along the ecliptic:

Pisces the Fishes
Aries the Ram
Taurus the Bull
Gemini the Twins
Cancer the Crab
Leo the Lion
Virgo the Virgin
Libra the Balance
Scorpio the Scorpion
Sagittarius the Archer
Capricorn the Goat
Aquarius the Water Carrier

For one-twelfth of the year the position of the sun moves slowly through one of these constellations and then it goes into the next constellation to the east. Master Vittorio could not resist telling Copernicus's explanation of the sun's slow motion through the ecliptic, though he carefully refrained from mentioning Copernicus's name. He merely said that the earth revolves around the sun in a little more than 365 days. The stars that nearly lie along a direct line from the earth to the sun when the earth is at one place in its orbit will obviously lie far from that direct line half a year later, when the earth has moved around to an opposite point on its orbit.

Master Vittorio explained these matters clearly, but there was one point that troubled the more thoughtful students. Finally one of them, a good friend of Tibaldo's named Stefano Costa, dared to raise an objection: "Sir, how do you know that the sun moves slowly among the constellations as the days of the year pass? When the sun is shining we don't see the stars, and when the stars are visible we don't see the sun." Master Vittorio was not at all displeased by Stefano's question. In fact, he was hoping that someone would ask it. He said, "That is a very good question! Can any one think of an answer?" No one answered immediately, but after some hints and reminders Stefano was able to suggest an answer. He recalled that when the class first viewed the evening sky in September they saw Scorpio setting soon after sunset. But a month later the constellation that set soon after sunset was Sagittarius, and a month later than that it was Capricorn that set soon after sunset. Obviously during those two months the sun had moved among the constellations. Master Vittorio was very pleased with this answer. He told the students, "You walk on the ground just like ordinary human beings, but you are beginning to have your minds in the heavens, like real astronomers."

Master Vittorio was so enthusiastic that he continued to tell other important consequences of the revolution of the earth around the sun. Because the axis around which the earth performs its daily rotation is tilted relative to its orbit of revolution around the sun, a point north or south of the equator of the earth is not exposed to the sun for the same amount of time each day. Consequently, everywhere except on the equator there is a cycle of seasons—from spring to summer to autumn to winter and then back again to spring—as the earth goes around the sun. The durations of daylight and of night are equal twice a year, at times called the vernal equinox and the autumnal equinox. The vernal equinox occurs in the northern hemisphere on a certain day when the sun is in the constellation Pisces. The day of longest daylight and shortest night, called the summer solstice, occurs a quarter of the year after the vernal equinox. And the day of shortest daylight and longest night, called the winter solstice, occurs a quarter of a year after the autumnal equinox.

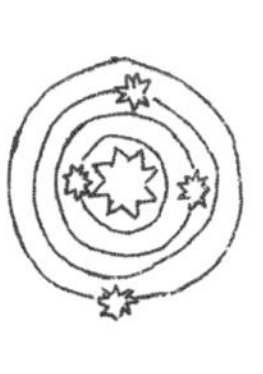

For a long time after Master Vittorio's explanation none of the students said anything, because they were full of wonder at the steady revolution of the earth around the sun, regularly changing the seasons from spring to summer to autumn to winter and finally back to spring again. But there was something in Master Vittorio's lecture that reminded Tibaldo of things he heard had in his visits to the Medical School of the University of Bologna. The constellations that the sun passed through in its yearly circuit—Pisces, Aries, Taurus, Gemini, Cancer, Leo, Virgo, Libra, Scorpio, Sagittarius, Capricorn, and Aquarius—were exactly the ones the astrologers considered important for shaping the life of an individual human being. He had overheard medical doctors solemnly discussing what is the best herb for treating swollen legs if the patient was born under the constellation Aquarius, which is certainly different from the most appropriate herb if the patient was born under Libra. Therefore Tibaldo could not refrain from asking, "These constellations the sun passes through along the ecliptic, do they determine a person's health and fortune as the astrologers say?" Master Vittorio really did not want to answer this question. He himself did not believe in astrology, but he knew that Master Domenico, who had hired him, was enthusiastic about the subject, and Vittorio did not want to risk losing his teaching position. Therefore he avoided the question and said, "Of course the motions of the heavenly bodies influence human life. As the sun passes through the constellations of the ecliptic the seasons change from spring to summer to autumn to winter, and our lives are very much shaped by the seasons." But Tibaldo was insistent. He said, "Yes, Master, I know that. But the march of the seasons affects everybody in the same way, and what I want to know is whether the patterns of the stars and the planets at the birth of a person affects that person in an individual way." Tibaldo had set a trap for Master Vittorio, who did not want to fall into it and so pretended ignorance. He answered, "I am just a humble astronomer, who knows a little about the motions of the moon, the sun, the planets, and the stars, But I know nothing about strange forces they may exert upon people. The only thing I can do is tell you my own case, which may not be typical at all. My parents have told me that at the moment of my birth the planets Mars and Venus were close together—in conjunction, as the astronomers say. Since the Roman god after whom the planet Mars is named was the strong god of war, and the goddess Venus was the goddess of love, my parents said that I would be very lucky at love. Well, I still don't have a wife or even a girl friend, and I cannot help thinking that if my salary were doubled it would do more good for my luck at love than the conjunction of Mars and Venus at my birth. But don't pay any attention to my case, because it may be a rare exception."

The Trouble with the Calendar

In January of 1582 there was a serious riot in the streets of Bologna because of a rumor that the calendar would be changed. When changes in the calendar were indeed imposed in Catholic countries in February of 1582 there were disturbances in many cities. Ten days were omitted when the old calendar was changed to the new one and the rioters shouted, "Give us back our ten days!" About a century and a half later, when the calendar was changed in Protestant countries, the pattern of unhappiness and rioting was repeated but with some new grievances. A larger correction was needed because of longer accumulation of error, and religious conflicts complicated the question of reforming the calendar. Now the rioters shouted "Give us back the eleven days the Pope and the Devil have taken away!" People generally like to keep what is familiar to them, but changes of the calendar caused special fears—that the crops would not ripen, that birds would be confused, that the motions of the heavenly bodies would become disorderly, that people's lives would be shortened, and that religious observances would be profaned. To understand these strange disturbances we must understand how the old calendar was defective and what changes were made to correct its defects. And in order to understand these things we can use some of Master Vittorio's lessons about the motions in the heavens that determine the day and the year. There are many motions in nature that repeat over and over again, with very little difference from one repetition to another. Each such motion is a natural timekeeper, and the length of each of its repetitions can serve as a unit of time.

The solar day is an example of a unit of time that is very important to human beings. Between sunrise and sunset the sun seems to move in a great arc in the sky. At one moment during the sun's course from sunrise to sunset, the sun reaches the highest point in this arc. We can call this moment "noon" if we agree to define that word by the arc the sun traverses and not by twelve o'clock on a man-made clock. Even though the duration of daylight between sunrise and sunset varies from one season to another, and hence the time interval between one sunrise and the next is variable, the time interval between one noon and the next is almost constant throughout the year. This time interval is the day, or more accurately, the solar day. We can measure a stretch of time by counting how many noons have passed since an initial noon. Fractions of a day are harder to measure, but they can be estimated by observing how far the sun or the stars have progressed in their arcs through the sky. Of course, the invention of mechanical clocks provided an accurate way to measure fractions of the day.

Another motion that repeats over and over again is the sun going around the constellations of the ecliptic. If we start to observe the sun when it is at a particular point in a particular constellation, for example at the beginning of Pisces, then one year has passed when it returns exactly to this point. Long intervals of time can be measured by counting how many years have passed since the beginning of the interval.

There are other natural units of time, such as the lunar month, which is the time required for the moon to go through all its phases from a new moon to the next occurrence of a new moon. The day and the year, however, are the most important units for organizing human life. The alternation of daylight and night during the period of one day determines the pattern of waking and sleeping for most people. The succession of seasons during one year governs the planting and harvesting of crops.

Since we have such good natural time-keepers, why do we need a calendar? The answer lies in the fact that human beings are social animals. We live not only in the great world of nature but also in the middle-sized world of human society. Calendars are needed to coordinate social activities. An election, for example, is a social activity. The voters must know when to come to voting places, with the assurance that ballots and boxes for voting will be provided. Likewise, a fair is a social activity. For example, the great gastronomical fair of Dijon, France, is held on the weekend closest to November 11. A farmer does not want to travel to Dijon with his sausages, herbs, pots of mustard, and other delicacies to find that there is no assembly of buyers for his goods, and the people who buy want to be sure they will find marketing booths when they arrive. A calendar is needed to ensure that buyers and sellers arrive at the fair at the same time. Religious holidays are social activities, and there are two reasons why they should be celebrated at the same time by adherents of a particular religion. First, when people with similar beliefs gather together to perform ceremonies sacred to them all, they feel a sense of community. Second, many religious holidays are anniversaries of events that occurred in the past at certain seasons, and their significance is lost if they are not celebrated at the same times of year as the original events.

Tibaldo had his own reason for being interested in the calendar. It was the calendar that determined when he could visit his own home. Tibaldo certainly did not want to leave the school of St.-Joseph-in-the-Corner, for he was ambitious to become a physician, and the school was a necessary step toward his goal. But the fact of the matter was that Tibaldo often suffered from homesickness. He was, after all, only a young boy—seven years old when he started at the school and eleven years old in 1582, when his strangest adventure occurred. He loved his crowded home with his parents, brothers, sisters, and nephews. It was a place where he was treated as an important person, whereas at school he was just one student among many. He looked forward to the times when he could go home. On Sunday he was allowed to spend the afternoon and evening with his family. On a few religious holidays—All Saints Day, Christmas, Easter, and Ascension Day—the school was closed and all the students were sent home. There was also one month of vacation in the middle of summer. Finally, Tibaldo had special permission from Master Domenico to return to the Bondi household to celebrate his birthday. He had begged Dr. Turisanus to request this special privilege from Master Domenico. As we mentioned earlier, birthdays were magnificent occasions in the Bondi household, and it was essential for Tibaldo to be king for the day in his own home on his birthday. Since Tibaldo treasured his time at home, he paid close attention to a tablet of the calendar that hung in the hallway of his school. He was soon to find, however, that there were some troubles with the calendar, more dangerous than he could have imagined.

Why should it be difficult to arrange the days of the year into a calendar? All that is needed, one might think, is to choose a first day of the year and then count the solar days until one solar year has elapsed. If one wants subdivisions of the year that are longer than a day, one can use another natural time-keeper, the moon, to define months, or introduce a conventional grouping of days like a week. Unfortunately, nature does not cooperate with all of these suggestions. Most troublesome, there is not a whole number of solar days in one solar year. In fact, the length of the solar year is 365.2422 solar days, or 365 days, 5 hours, 48 minutes, and 46 seconds. If, for example, a new year begins at midnight, and one wants the next new year to begin when exactly one solar year has elapsed, then the next year would begin at 5 hours,48 minutes, and 46 seconds after midnight. And then the following new year would begin 11 hours, 37 minutes, and 32 seconds after midnight; and so on and on. That would obviously be a messy way to arrange a calendar. A useful calendar must have a whole number of days in each year, and only on the average may the length of the year be 365.2422 solar days. As to dividing the year into months, the period from one new moon to the next is 29 solar days, 12 hours, 44 minutes, and 3 seconds, and therefore there are 12.37 of these lunar periods in one solar year. Consequently, if one wants twelve calendar months in the year, and each month is to have a whole number of days, it is necessary to make the calendar months have unequal numbers of days.

The calendar used in most of the countries around the Mediterranean and most of Europe for more than sixteen hundred years was the Julian calendar. It was named after Julius Caesar, who established it in the year 45 BC, when he was ruler of Rome and hence of the large empire around the Mediterranean that the city of Rome had conquered.

There had been a calendar in Rome before Julius Caesar, but it was a bad one because its rules were indefinite. The priests who conducted the rituals of the Roman gods and goddesses were entrusted with adjusting the lengths of the calendar months to make the calendar year agree approximately with the solar year. They did so corruptly by adding days in such a way as to lengthen the terms of office of officials whom they favored.

The Julian calendar was an improvement because it

had definite rules and was simple. Following the advice of astronomers of his time, Julius Caesar assumed that the solar year was exactly 365.25 solar days long. To make the calendar year have this length on average, he commanded that three years out of four would be 365 days long and the fourth would be a leap year, 366 days long, an extra day being inserted at the end of February. He did not know that the true length of the solar year is 365.2422 solar days, and therefore that his calendar year was on average 11 minutes 14 seconds longer than the solar year. Had Julius Caesar known about the inaccuracy of the length of his calendar year, would he have been troubled? Of course we do not know, but somehow it is hard to imagine that the strong-willed, self-centered general who had conquered Gaul, invaded Britain, and defeated his Roman rival Pompey in a bloody civil war would have worried much about a deviation of 11 minutes 14 seconds per year. There were, however, people living long after Julius Caesar who did worry about this deviation. Caesar would have been profoundly baffled by their reasons for worrying and even more baffled that they had sufficient power to treat their worry by changing his calendar—a calendar of which he was very proud, as he was of everything he did. The one thing Julius Caesar would have understood about Gregory, the man most responsible for changing the calendar, was that he lived in Rome. But that Gregory would rule in Rome not because he was a military conqueror but a priest—and a priest of a religion that did not even exist in Julius Caesar's time—Ceasar would have found incomprehensible.

Between the death of Julius Caesar in 44 BC and the reform of the calendar by Pope Gregory in 1582, there occurred one of the most important revolutions in human history: the beginning of Christianity and its spread until it became the dominant religion of Europe. The details of this revolution would lead us far from our story, but we cannot neglect the main facts. Indeed, the little world of Tibaldo was part of the middle-sized social world in which Christianity was the central feature. Even when we use the date 1582 we are unconsciously recognizing the immense influence of Christianity, because dates are counted as the number of years that have elapsed since the birth of Jesus. According to Christians, Jesus was the savior whose coming was predicted by certain Jewish prophets, a savior called by them *Messiah,* the Hebrew word for *anointed one.* The Greek word for *anointed one* is *Christ,* and that is why the followers of Jesus called themselves Christians and called their religion Christianity. As this religion developed, Jesus was accepted by Christians not only as a human savior but as a divine person.

Jesus was born in Judaea, a country that had been conquered by a Roman general not long before Julius Caesar became ruler of Rome. The inhabitants of Judaea, the Jews, were unwilling subjects of the Roman Empire, especially because some of the demands made upon them, such as worship of the emperor, were contrary to their religion. The Roman governors of Judaea were harsh in their treatment of rebels and put many to death by the horribly cruel method of crucifixion. Jesus himself was suspected of being a rebel and was crucified at the age of thirty-three. According to the Christians, Jesus rose from his grave on the third day after his crucifixion and ascended to heaven. That day of resurrection was a Sunday, and its anniversary is celebrated by Christians as Easter, the most important of all holy days of the year.

At first the Christians were a tiny sect among the Jews, but Saint Paul and other energetic missionaries began to gather converts from many peoples within the Roman Empire. Although the Christians were fiercely persecuted by the Roman government for more than two centuries they continued remarkably to increase in number. A great turning point was the conversion of the Emperor Constantine to Christianity in the year 312. Although he tolerated other religions, including the ancient religion of Rome with its many gods and goddesses, Christianity was favored and soon became the official religion of the Roman Empire.

Constantine was disturbed that there were many disagreements in beliefs among the Christian leaders. If the leaders did not agree, what should ordinary people believe? He therefore summoned in the year 325 a great meeting, or Council, in the city of Nicaea, in what is now Turkey, not far from Constantinople, the city that he had made the capital of the eastern half of the Roman Empire and had named after himself. The Emperor Constantine himself presided over the Council and made sure that almost unanimous agreement was reached on a set of beliefs—called the Nicene Creed—which Christians were then obliged to accept.

One of the accomplishments of the Council of Nicaea was devising a formula for setting the date of Easter. Easter was a spring holiday for two reasons. The first is that the Jewish holiday of Passover occurs in the month containing the vernal equinox, and the Last Supper, an event of great significance in the life of Jesus just before his crucifixion, was actually a Passover celebration. The second is that the very idea of resurrection is mixed up in people's minds with springtime, when plant life revives on earth. To ensure that Easter would remain a spring holiday, the following rules were adopted at the

Council of Nicaea: (1) the date March 21 in the Julian calendar was declared to be the date of the vernal equinox; (2) Easter was to be celebrated on the first Sunday after the first full moon that occurred on or after the vernal equinox.

This formula for setting the date of Easter is definite and fairly simple. It has, however, several flaws. The worst flaw is that declaring March 21 to be the date of the vernal equinox does not make it so. The interval between one vernal equinox and the next is one solar year. Because the calendar year in the Julian calendar is 365 days long three years out of four, but 366 days long on the leap year, it is impossible for the true vernal equinox to fall on March 21 every year. That would not be serious, however, if the true date of the vernal equinox merely danced among March 19, March 20 and March 21, because if that were so the formula at Nicaea would still make Easter a spring holiday. The serious trouble is that the vernal equinox steadily moved to earlier and earlier dates. On average the calendar year according to the Julian calendar is 11 minutes and 14 seconds longer than the solar year, and that deviation accumulates as the years pass by. In 128 years the accumulated deviation amounts to one whole day, and by 1581, 1256 years after the Council of Nicaea, the deviation amounted to ten days. By then the true vernal equinox was occurring around March 11. Consequently, calculating the date of Easter on the assumption that the vernal equinox occurs on March 21 had the effect of moving

Easter ten days toward summer. By the year 10,000, Easter would be occurring close to the summer solstice. Master Vittorio had already pointed out this flaw to his class and had commented that no one would do anything about it because the astronomers who understood the problem did not have the authority to reform the calendar. For once Master Vittorio was wrong. There was someone who understood the problem without being an astronomer, and he did have authority.

That person was the Bishop of Rome, called the Pope (from the late Latin word *papa,* meaning *father*). He was a native of Bologna, given the name Ugo Buoncompagni at birth. As Pope he was called Gregory XIII, and he ruled from 1572 to 1585. By the late Middle Ages the bishop of Rome had become recognized in western Europe as the head of the Catholic Church, with the power to appoint other officials of the Church and to

regulate the beliefs and practices of members of the Church. The word catholic means universal in Greek, but that name was not exact, because the Christians of Asia and eastern Europe recognized the authority of several patriarchs, especially the patriarch of Constantinople, rather than the authority of the Pope. Furthermore, beginning in 1517 the Protestant Reformation spread through much of northern Europe, and all the different Protestant churches agreed in rejecting the authority of the Pope.

Nevertheless, the Pope's authority was immense, continuing in Spain, Portugal, France, Italy, Austria, Hungary, and much of Germany, as well as the vast colonies in the Americas that belonged to the first three of these countries. In addition to being the head of the Catholic Church, the Pope up until the middle of the nineteenth century ruled an extensive territory in Italy stretching far

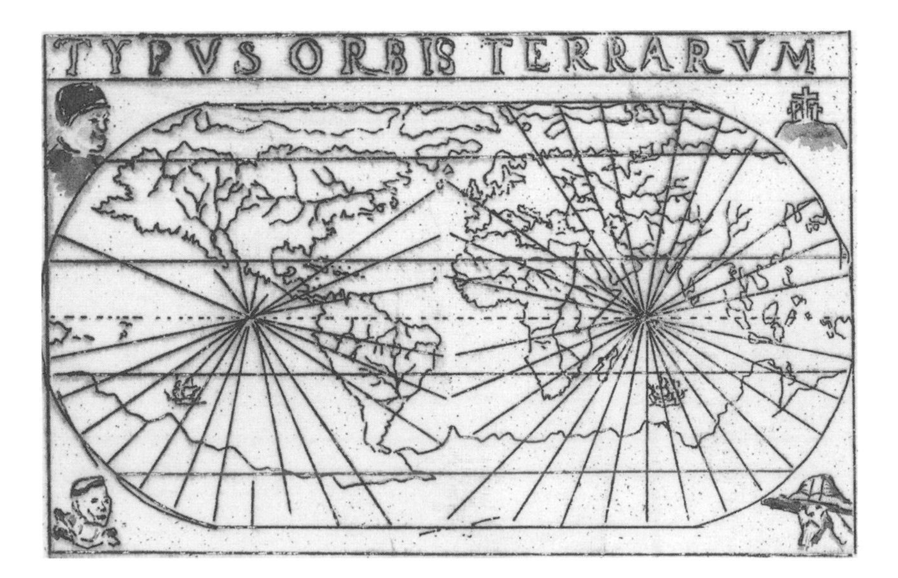

beyond the city of Rome. He collected taxes, maintained a police force, built roads and other public works, and raised an army (and occasionally fought battles using it). For example, the city of Bologna, where Tibaldo lived, was part of the territory of the Pope, even though it was more than three hundred kilometers from Rome.

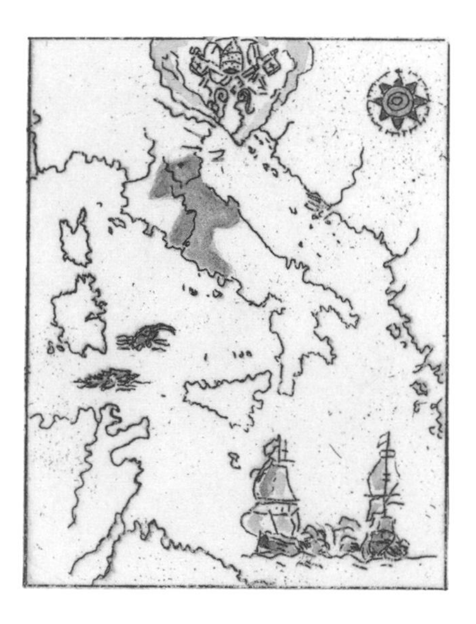

Gregory not only understood the flaws of the Julian calendar but realized that a decree to reform the calendar would be a dramatic exhibition of his authority. Furthermore, he hoped it would be a popular act that would be useful in his campaign against Protestantism. Consequently, in 1577 Pope Gregory appointed a Commission, including the great German mathematician and astronomer Christopher Clavius, for the purpose of designing an improved calendar.

The Hole in the Calendar

The great problem the Commission faced was to design a calendar accurate enough for all practical purposes but not so complicated that only a mathematician could use it. The calendar of Julius Caesar, with its rule for adding an extra day to February every fourth year, is wonderfully simple but inaccurate, as we have seen. Some astronomers recommended rules that were so accurate that the accumulated error in a million years would be no more than one day, but no village priest, trying to figure out when Easter would fall each year, would be able to perform the calculations required by those rules. A critic of those complicated rules remarked, "Easter is a festival, not a planet."

The Commission decided on the following rules. Years not divisible by four should have 365 days. If a year is divisible by four but not by one hundred, it is to be a leap year, 366 days long, with the extra day added to February. If a year is divisible by one hundred but not by four hundred, it would have only 365 days. Finally, if a year is divisible by 400, then it would be a leap year. Therefore, 1584, 1588, 1592, 1596, 1604, etc. would be leap years; 1700, 1800, and 1900 would not be leap years.; but, 1600 and 2000 would be. These rules are a compromise between simplicity and accuracy. Although they are more complicated than the rules of the Julian calendar, they are not too difficult for most people to master. As to accuracy, they would cause an error of 2 days, 14 hours and 24 minutes in 10,000 years. In 40,000 years, therefore, there would be an accumulated error of 10.4 days, which is approximately the deviation that Gregory XIII wanted to correct. Perhaps in the year 41,582 Pope Gregory CXIII will summon a new Commission for a new reform of the calendar. But that bridge can be crossed when we come to it.

The Commission had to face one further problem. What should be done with the error that had accumulated in the Julian calendar since the year 325, which had the effect of making the true vernal equinox fall on March 11, instead of on March 21 as decreed at the Council of Nicaea? Obviously some dates simply had to be skipped. But which ones? And what would be the consequences of skipping dates? Even before the Commission made its report to Pope Gregory there were rumors flying around that dates would be omitted from the calendar during the year 1582. After all, when any group of people with twelve or so members meets to decide some matter of public importance, it is almost inevitable that some information will leak out concerning the deliberations of the group, and little leaks cause large rumors. There were rumors that people who rented houses would have to pay rent for a whole month, but the month from which days were omitted would only have twenty days. And there were rumors that loans would have to be repaid ten days earlier than would have been required if the Julian calendar had been kept. And there were rumors that the crops would spoil because they would not know when to sprout in the earth. And rumors that birds would not know when to start migrating south in the autumn.

On January 7, 1582, there was a partial eclipse of the sun visible in Bologna. Master Vittorio had told his students to expect this event, and he warned them not to gaze directly at the sun, since they could injure their eyes, but to look at a reflection of it on paper. It was not a total eclipse, in which the moon entirely blocks the sun for a few minutes and allows people to see the wonderful glowing corona of the sun's outer rim. Rather, the moon first blocked a tiny piece of the sun, then a larger and larger piece, until about nine-tenths of the sun was blocked, after which the moon's motion gradually uncovered the sun. The entire process took about two and a half hours, and at the midpoint, when the sun was most blocked, the sky seemed strangely dim. Soon after the beginning of the eclipse the news of it spread rapidly through the city, and fear increased as more and more of the sun was blocked. Some people fled into their houses, bolted their doors, and shuttered their windows. Many went into the churches to pray. The streets and shops were mainly deserted. The chief of police, Captain Arcangelo, reported at the end of the eclipse to Signor Antonio Domitiani, the governor of Bologna, according to his usual procedure when a crisis occurred in the city. The governor asked his usual questions.

"Any looting?"

The captain answered, "No, Your Excellency."

"Any fires in the streets?"

"No, Your Excellency."

"Any suspicious characters?"

"Only the usual ones, Your Excellency."

"What did people do?"

"They stopped working, Your Excellency. Many prayed."

"Well, that won't do too much harm."

The two experienced men agreed that it had been a mild crisis, not comparable to the total eclipse of 1572.

The relief of Captain Arcangelo and Signor Domitiani was premature, however. The following day the captain reported unexpectedly and in considerable agitation to the governor.

"Your Excellency, il Torrentino has flooded."

You might suppose that the captain was reporting damage done by an overflowing river, swollen by the winter rains, since the Italian word *torrentino* means *small torrent*. As a matter of fact, the captain was referring to a hermit named Fra Zaccaria, who mostly lived alone inside a deserted tomb in the ancient Etruscan cemetery of Felsina, to the west of Bologna. He occasionally entered the city in great excitement and gathered a crowd to whom he preached about some evil that troubled him. He was like a little river that was dry during most of the year but overflowed during the rainy season, sometimes doing serious damage. Some of the poor people of Bologna thought he was a holy man and listened attentively to his sermons, while many others had doubts about his holiness but were afraid of his curses and his *malocchio* (evil eye). The governor asked,

"What is il Torrentino saying?"

The captain replied, "He is shouting about the eclipse."

"That can cause no trouble," the governor said, "since the eclipse is past and has hurt no one."

"He says that it has hurt no one yet but is a sign of terrible things to come."

"What things?" asked the governor.

"He says if the calendar is changed, as people say it will be, the heavens will be torn apart and the sun will be swallowed up permanently. The eclipse was sent as a warning to leave the old calendar alone."

The governor asked why il Torrentino thought that changing the dates assigned to days could do any damage to the heavens.

The captain answered, "He says that ten days are going to be omitted from the calendar, and therefore these days are going to be destroyed even though they have committed no sins."

"What does the crowd say?" asked the governor.

"The crowd says that ten innocent days are going to be destroyed."

"Why should they care?" the governer asked.

"Il Torrentino tells them that they will lose ten days of their lives."

"And what do they say to that?"

"They say, 'give us back the ten days of our lives.'"

The governor then asked whether il Torrentino was making any threats against Pope Gregory, who had appointed the Commission to reform the calendar.

The captain answered, "Il Torrentino says that Pope Gregory is an innocent but simple Pope, who is given wicked advice by his advisers, especially the German mathematician Christopher Clavius, a servant of the devil

who does black magic with numbers and pentagrams, and whose name spelled backwards is 'Suivalc,' which is one of the six hundred and sixty-six hidden names of Satan."

The governor asked, "What does the crowd say?'

"They say, 'Destroy Suivalc and all his devilish writings.' And they plan to march to the bookstore of the University of Bologna to burn all the writings on mathematics and astronomy."

Signor Domitiani said, "Il Torrentino and his crowd can make all the noise they want, but they are not to destroy property. Order your men to disperse the crowd. And politely invite il Torrentino to come to my office, where he will be given a chance to persuade me." He added, with a certain amount of self-satisfaction, "A governor who does not know how to handle the likes of il Torrentino does not deserve to be governor."

Signor Domitiani prepared for his visitor by instructing his servants to prepare an altar in his office, surrounded by statues of the Roman deities Jupiter, Mars, Venus and Mercury, with a bust of Julius Caesar among them and a tablet of the Julian calendar for 1582 lying on the altar. When il Torrentino entered his office, Governor Domitiani welcomed him with great respect. Then he said, "Fra Zaccaria, it is rare that I have the privilege to receive a man as learned and holy as yourself in my humble office. Before we commence any discussions of business, would you say a prayer for my soul and for the city of Bologna?" Then he led the hermit to the improvised altar. Il Torrentino did not ordinarily lose his power of speech, but he was dumbfounded by the pagan altar

that had been prepared for him. Finally, he exclaimed, "These are abominations. Graven idols and false gods. You will be damned eternally for this diabolical altar. And as for me—I would rather be burned at the stake than utter a prayer in this vile place."

The governor was not as easily frightened as the beggars of Bologna by il Torrentino's curses, and he said calmly, "But Reverend Father, if you look on the altar you will see a tablet of the Julian calendar, the sacredness of which you were just now preaching in front of the Palazzo Communale. And there is a bust of the Blessed Julius Caesar, who decreed the sacred Julian calendar. You should find this altar congenial for your prayers."

Now Fra Zaccaria really was at a loss for words, and Signor Domitiani made use of the interval of unusual silence to give his guest a lecture on the inaccuracy of the Julian calendar, the slipping of the date assigned to the vernal equinox, and the gradual drift of Easter toward the summer solstice even though religion requires it to be a spring festival. He concluded, "The new calendar that Pope Gregory, our Holy Father, will soon proclaim is a true Christian calendar, while the old calendar of Julius Caesar is a relic of paganism."

Fra Zaccaria was an excitable man, but he was not as mad as some believed him to be. He could see the strength of Signor Domitiani's arguments, and he also was aware that Signor Domitiani had less polite and less pleasant ways of dealing with his opponents than today's treatment.

In short, although he had entered Bologna that morning as il Torrentino, he left that afternoon as quiet Fra Zaccaria, a convert to the Gregorian calendar.

After he left, Signor Domitiani said to his assistants, "He'll come to Bologna again next year or the year after, perhaps preaching against the thieving tricks of merchants from Florence, the Babylon on the Arno, or the traders from Ravenna, the Sodom on the Adriatic. Il Torrentino can sometimes be useful when he floods."

By January 1582 Tibaldo had become aware of the rumors that the calendar might be reformed. He had heard long ago from Master Vittorio of the flaws in the Julian calendar, and recently Vittorio had admitted that he had made a mistake in prediciting that no one in authority would try to correct these flaws. Nor did the prospect of a change in the calendar trouble Tibaldo in the least. He understood that a change of the calendar would not make any difference to the order of nature, because calendars are only human instruments for keeping track of days. At age eleven Tibaldo was already a little conceited about his knowledge, and he looked down on il Torrentino and his followers as mere ignoramuses. What could be more stupid, thought Tibaldo, than believing that the numbers assigned to days of the year were designated by nature, and that the heavens would be damaged by skipping ten numbers in a calendar? Tibaldo felt superior to such silly beliefs.

It was on February 24, 1582, that Pope Gregory XIII proclaimed the new calendar. He imposed the rules for calculating leap years that his Commission had recommended, and he accepted the recommendation that ten days be omitted to correct the accumulated error of the Julian calendar. Specifically, he ordered that the day following October 4, 1582, should be designated as October 15, and the dates from October 5 through October 14 should not occur in 1582. The result would be that in 1583 the vernal equinox would occur on March 21 and for three thousand years thereafter would never drift more than two days from March 21. The first two weeks of October were chosen for skipping days because no important holidays of the Church occurred at that time.

Messengers were sent out from the Vatican Palace in Rome carrying the proclamation about the calendar, and in a few days the news came to Bologna. It reached Tibaldo and his classmates in an unusual manner. The teacher who supervised the translation of Latin into Italian liked to surprise his students with pages they did not expect and could not prepare in advance. What better surprise than a page of Latin that had just been written, as fresh as this morning's newspaper is today? Hence when the teacher, walking from his home to the school of St.-Joseph-in-the-Corner, saw a messenger posting Pope Gregory's proclamation in front of the Palazzo Communale, he copied it to use in his morning class.

The translation exercise began with the magnificent first words of Pope Gregory's proclamation *"Inter gravissimas pastoralis officii nostri curas...* ("Among the most serious responsibilities of our office as pastor ...") and continued with the Pope's review of the flaws of the Julian calendar and his statement of the rules of the new calendar. As always, two dozen students began writing Italian translations on their slates, among them Tibaldo. After four years in Master Domenico's school he translated automatically, one language going in at the ears and another coming out at the fingers; the mind hardly became involved at all. After half an hour the dictation was completed. The teacher began walking up and down the aisles, checking the accuracy of the translations while twenty-four boys were squeezing the cramps out of fingers that had written too fast.

While Tibaldo was waiting for the teacher to examine his slate, he looked over what he had written. Most of it did not surprise him. He had learned from Master Vittorio about the defects of the old calendar, and he had heard rumors of the recommendations of Pope Gregory's Commission. One of the debates at the school of St.-Joseph-in-the-Corner, which Master Domenico liked to arrange, concerned the desirability of a new calendar, and Tibaldo had argued cleverly and convincingly in favor of it.

What was new to Tibaldo, and what took him by surprise, was that all the days from October 5 until October 15 would be omitted in 1582. There would be a hole in the calendar! And the most important thing that would fall through the hole in the calendar was October 10, Tibaldo's own birthday!

He had already had his eleventh birthday in 1581, and he would have his thirteenth birthday in 1583, but his twelfth birthday would never occur! The hole in the calendar meant that there would be a hole in his life!

Without thinking, Tibaldo exclaimed aloud, "I shall lose my birthday!" The teacher wheeled around in the aisle and faced him, asking, "Bondi, what did you just now say?" Speaking out in class without permission was strictly forbidden, and the period when translations were checked was supposed to be the quietest of all. Tibaldo came to his senses and started to apologize, but he did not explain why he had cried out. He knew that his teacher was so dull and dry he would never sympathize with the true explanation. The teacher did not accept Tibaldo's apology but made him stretch out his hands to receive a blow from the birch switch he always carried. Tibaldo suppressed his tears and said nothing. He was angry at the punishment, however, and his anger became mixed up with his unhappiness at the prospect of losing his birthday. Each pain made the other hurt more.

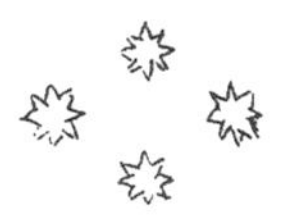

Tibaldo Begins the Battle

Tibaldo stayed at the school of St.-Joseph-in-the-Corner day and night Monday through Saturday, but at Sunday noon he was allowed to return to his parents' home until early Monday morning. The first Sunday after he had translated Pope Gregory's proclamation he explained his misery to his father and mother.

His father, Lorenzo Bondi, said to him, "The time has come to grow up and forget childish things like birthdays. Even though you are the son of a poor man you have been given an opportunity to obtain an education and become a doctor of medicine. You should keep your mind on your schooling and your future and not be distracted by childish things." Tibaldo was not able to reply to his father, because he deeply respected Lorenzo and knew that what he had said was correct—or at least partly correct. Tibaldo was a little ashamed of his own childishness, but at the same time he felt that it was not wrong for an eleven-year-old boy to love some childish things, no matter what he would become in future years. But all his training in debating did not equip him to explain these complicated feelings to his father.

Tibaldo's mother, Teresa, tried to comfort him by promising that on the day of his birthday according to the old calendar, October 10, the Bondi family would celebrate as gloriously as ever with all the usual special delicacies for dinner, and Tibaldo would be king for the day. Tibaldo refused to be comforted. He said, "I shall have no official birthday in 1582. The old calendar will no longer exist, and therefore October 10 will not exist. And according to the new calendar my birthday will not exist because there will be a jump from October 4 to October 15. If we have a birthday dinner for me in our family, it will be counterfeit, not the real thing." Tibaldo was vaguely aware that he was being childish and peevish by arguing with his mother in this way. After all, he had learned from Master Vittorio that the passage of a solar year between one day and another has nothing to do with the numbers assigned to those days but is a matter only of the completion of a cycle in the motion of the sun through the ecliptic. It is the passage of that much time that would make him a year older, no matter what calendar one used. In other words, Tibaldo was behaving like the ignorant followers of il Torrentino, with their superstition that nature assigned definite numbers to the days. Tibaldo was a little ashamed of himself for behaving as if he had never studied astronomy, but he also learned an important lesson about himself. He realized that he did not differ as much he had thought from uneducated people.

Since Tibaldo was not satisfied with the advice he had received from his parents, he decided to ask a teacher whom he admired and trusted at school. There were two possibilities—Master Demetrios and Master Vittorio. Tibaldo knew that it was hopeless to ask Master Vittorio, who would just tell him to study nature and not be trapped by ignorant superstitions. He decided, therefore, to consult Master Demetrios, who he hoped would be tolerant of his obsession and would have some useful suggestions of strategy. We must report that Master Demetrios did not take the matter as seriously as Tibaldo. In fact, when he understood what troubled Tibaldo he began to laugh. And he gave a little lecture because it was part of his nature to give lectures: "Bondi, is it such a serious thing to lose a birthday? Ought you to be unhappy about such a triviality? Haven't I told you that you must aim at great deeds and not be disturbed by the little annoyances of ordinary life?"

One part of Tibaldo agreed with Master Demetrios, but another part of him continued to feel that a birthday is too precious to lose without regret. He also noticed that even though Master Demetrios laughed at him it was not a harsh and mocking laugh. It was a good-natured laugh of a man who still had some sympathy with the concerns of a small boy. Hence Tibaldo had the confidence to answer, "Yes, Master, my birthday is no great thing. But I am still too young for great things. Perhaps some day when I grow up I shall join an expedition to recapture Constantinople, or I shall find a remedy for the plague. But now I am little and can only do the little things for which I am fit. If I do little things well, then I shall be trained to do great things later. And now I intend to recover my birthday." Four years of practice in debating were turning out to be useful for Tibaldo.

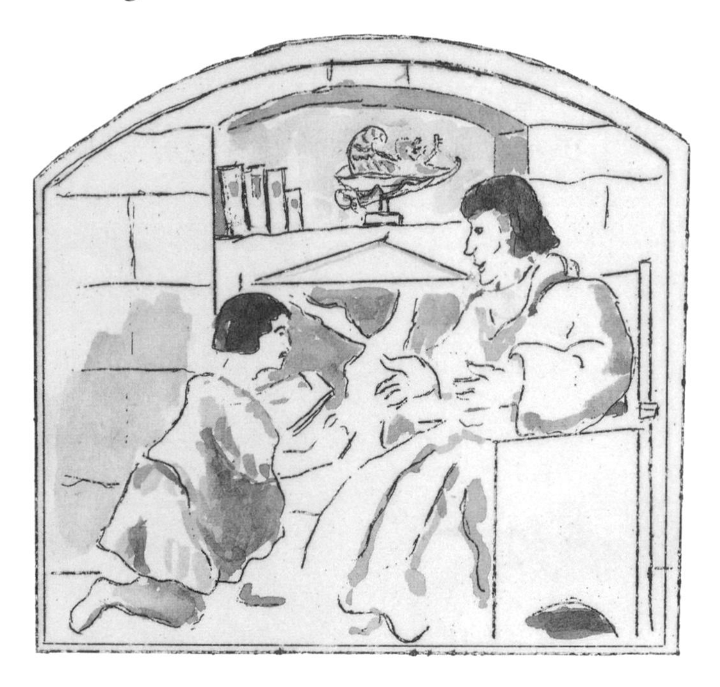

Again Master Demetrios laughed heartily. He said, "I can't give you your birthday, Tibaldo, because I don't own it. In fact, I have very few possessions. If you look in my cupboard you'll find some old clothes, a Greek grammar book, a copy of Homer's *Odyssey,* and a map of Constantinople that my great-grandfather saved when he fled the Turks and came to Italy. But I can give you something that a poor teacher has plenty of, and that is advice. My advice is that you take your problem to a man of power in Bologna. There really is only one—the governor, Signor Antonio Domitiani. You cannot go to him directly, because he will not receive you. If you know some one who knows him, however, or some one who knows some one who knows him, then perhaps you'll be able to present your case to the powerful gentleman."

Now Signor Domitiani was not a king, nor a prince, nor a duke. As we saw earlier when Signor Domitiani dealt with Il Torrentino, he was the governor, and he governed in the name of Pope Gregory XIII. Bologna was not an independent city, like Florence or Venice or Milan, but belonged to the part of Italy known as the Papal States. As we explained before, the Pope ruled in two very different ways. He was head of the Catholic Church, and he also controlled the military, economic, and political affaris of about one-tenth of the territory of Italy, as if he were a prince. Pope Gregory was so busy with his many duties, however, that he needed governors for the individual cities and districts of the Papal States. Signor Domitiani was the pope's reliable governor of Bologna and the surrounding countryside. It was Signor Domitiani who made nearly all of the decisions concerning Bologna, and the Pope almost never overruled him. For this reason Master Demetrios advised Tibaldo to take his problem to Signor Domitiani.

Since Tibaldo was an eleven year old boy and the son of a poor man with no influence, one might think that Master Demetrios's advice was useless. Was not Signor

Antonio Domitiani as remote from Tibaldo Bondi as the planet Saturn from the earth? But no, the advice was not useless and the situation not hopeless. Tibaldo knew that Professor Turisanus, the famous man who paid his expenses at school and had in a way adopted him, was the personal physician of Signor Domitiani. Once a week the Professor examined Signor Domitiani, with more frequent visits in case of illness.

Tibaldo decided to ask Professor Turisanus to take him along on the occasion of the next examination. Tibaldo would wheedle the Professor by saying that the only way to become a medical doctor is to see how a great doctor actually works. Tibaldo knew that this strategy had a chance of succeeding, but care was necessary in one thing. Tibaldo's father was always afraid that Professor Turisanus might become annoyed with Tibaldo and stop paying for his schooling. Consequently, Tibaldo was forbidden to ask the Professor for any new favors. It was a hard decision for Tibaldo. Ordinarily he was an obedient son, but the present situation was special. The likelihood that Tibaldo would lose his birthday had become an obsession with him. As a result, he was disobedient for once in his life. He wrote a wheedling letter to Professor Turisanus, obviously not mentioning his birthday and the hole in the calendar. He pretended only that he wanted to see a great physician at work. We cannot hide the fact that Tibaldo did not show good character on this occasion, but in our story we must present the facts as they occurred.

Professor Turisanus was so fond of Tibaldo and so easily flattered that he agreed to take the boy along on his next visit to Signor Domitiani. He sent a servant to the school with a note to Master Domenico, asking that Tibaldo be excused from afternoon classes. Master Domenico, who was always extremely respectful of influential men, permitted the servant to take Tibaldo from school. Together the Professor and the boy rode in a coach to the Governor's palace. Professor Turisanus was recognized by a series of guards carrying lances, then a series of guards carrying sabers, then a series of secretaries and clerks, and finally a series of assistants and officials. The two visitors were admitted though a series of armored gates and polished doors.

In an immense room with high pointed windows, beams along the ceiling, tapestries on the walls, and paintings of generals and cardinals in the corners was the Governor's massive, gleaming desk. However, the Governor was not seated at his desk in his usual official manner. Instead, he was stretched on a couch near the desk, twisting and turning with a terrible stomach cramp. Professor Turisanus immediately began an examination. He felt the Governor's pulse, looked into the pupils of his eyes, thumped on his chest, and then demanded that a servant nearby bring him a large brass bowl. From his medicine bag Professor Turisanus took a strange looking green flask, poured a few drops into a cup, and mixed in some red fluid from another flask. This mixture was an emetic of his own invention, designed to empty a patient's stomach. The Governor was asked to drink the contents of the cup, and indeed within two minutes the action of the emetic started. Soon the Governor's stomach was empty, and the brass bowl was full.

It was a most interesting medical lesson for Tibaldo. The servant carried off the bowl, but Professor Turisanus stayed by his patient's couch to supervise the recovery, and Tibaldo stayed with him.

Although Signor Domitiani was much relieved, he was exhausted from the unusual exertion. Little by little his strength came back, and he was in a wonderful mood because his stomachache was gone. He became curious about the intelligent looking boy who was watching intently and began to ask him questions. Tibaldo saw his chance. He cleverly brought the discussion around to Latin translation and then mentioned, as if it were an after-thought, the Pope's proclamation about the change of the calendar. The Governor sighed, "No end of trouble. Every one who works for the city of Bologna wants to be paid a month's wages for October, even though October of 1582 will contain only twenty-one days. When I tell them they will be paid two-thirds of a month's wages they feel cheated and say that a month is month, even if it has only twenty-one days. And besides, my clerks are too incompetent at arithmetic to calculate two-thirds of the monthly wages. If that isn't bad enough, a mad hermit rushed into Bologna stirring up the crowd to believe that the heavens will fall down because

ten days have been omitted from the calendar. No end of trouble, no end of trouble."

Tibaldo quickly took the opportunity to say what trouble the hole in the calendar was causing him—that he would lose his twelfth birthday. The Governor laughed and said, "Yes, you have your problem, and I have my problem, but mine is ten thousand times more complicated than yours." The quick-witted Tibaldo replied, "Yes, Your Excellency, but I can't do anything about it and you can, because you are very powerful." The Governor was amused again and answered. "No, young man. Some things I can do, and some things I cannot. The change in the calendar was ordered by Pope Gregory. whose servant I am. What he orders I carry out."

Tibaldo, still quick-witted, was about to suggest that the Governor appeal to the Pope to change his mind, but he noticed that the look of amusement was disappearing from the powerful man's face, like the sun behind a storm cloud. Signor Antonio Domitiani was beginning to have another stomach cramp. It was a sad fact that Professor Turisanus's remedies did not always have long effects. A new dose was given to the Governor, along with an additional potion to put him to sleep, and the Professor and the boy then left the magnificent room rather quickly, neither particularly satisfied.

Professor Turisanus was angry with Tibaldo for making his foolish request, and even began to blame him for the return of the Governor's stomach cramp. That was ridiculous, because the cramp had nothing to do with the conversation. Professor Turisanus, however, like many men whose work has not succeeded, wanted to blame some one other than himself. Tibaldo was really frightened. Never before had the Professor been angry with him. He feared that the old doctor would stop paying the expenses at St.-Joseph-in-the-Corner and that Tibaldo would have to quit his studies. And then his parents would be broken-hearted. All because of his persistent obsession with a birthday!

Fortunately for Tibaldo the Professor soon calmed down. He really was fond of the boy. Besides, he knew that he had invested four years in Tibaldo's education. If Professor Turisanus was to have a successor who would be a substitute son, there was no time to start again. It had to be Tibaldo or nobody, and therefore it was still Tibaldo.

The Birth of a Baby and a Plan

The following Sunday afternoon Tibaldo spent as usual in his parent's house, and of course he said nothing about the daring but unsuccessful visit to Signor Domitiani. He tried to keep a cheerful appearance that would cover up his deep sadness and disappointment. However, Tibaldo's oldest sister, Anna Maria, knew him well enough to recognize the sadness be-neath his forced cheerfulness. When she questioned him, he began to cry, which was embarrassing to him, since after all he was the uncle of Anna Maria's two children, and uncles are supposed to

be strong. Ti-baldo admitted to his sister that he was still brooding about the loss of his twelfth birthday.

Now Anna Maria was not the sort of person to waste much sympathy on Tibaldo's childish obsession. No one could be farther from childishness than she. She had nursed her husband in his illness and had experienced the sorrow of his death after only a few years of marriage.

She knew the responsibilities of working to support her children and of raising and educating them (fortunately with the help of her mother when her work called her away from home). Furthermore, Anna Maria's work was serious, strenuous, and full of risks. She was a professional midwife, who assisted women in childbirth. Often she was called in the middle of the night to assist a woman who had suddenly begun to have labor pains, which is the name for the pains that accompany muscular contractions as the baby is preparing to emerge. Anna Maria sometimes had to stay up without sleep for twenty hours or more when the birth was slow, and she had to be alert during that time to deal with emergencies. Furthermore, she needed to stay calm to influence the laboring woman to relax, since tension greatly increases the difficulty of child-bearing.

A risk that threatened all midwives was the possibility of being accused of witchcraft, because it was a superstitious age, and there was widespread belief in the existence and power of witches. It sometimes happened when a baby was dead at birth, or died soon afterward, that a superstitious and hysterical parent would accuse the midwife of causing the baby's death as an offering to the devil. That was, of course, a stupid accusation, and stupid reasons were given to support the accusation, such as strange sounds in the house or strange behavior of cats and farm animals. But you can imagine how difficult it is for a midwife to prove that she did not practice witchcraft when she is surrounded by people who interpret any unusual occurrence as the devil's work! But what if a midwife has a very good record in helping to deliver babies? You would think that then she is safe. But no, there were people who would accuse of her being a "white witch," not serving the devil, to be sure, but still making use of supernatural methods, which could be dangerous in ways that we do not understand. In other words, accusations could be made against a midwife whether she was successful or unsuccessful. Midwives were needed, however, to help not only poor women during childbirth, but also the wives of rich and influential citizens, since physicians did not believe this task was dignified enough for them. Consequently, rumors of witchcraft almost never led to arrests, trials, and punishments. Nevertheless, in many European countries the occurrence from time to time of terrible persecution of women suspected of witchcraft—even burning at the stake—was enough to make midwives worry when stupid rumors were spread about them.

Anna Maria loved her little brother dearly but seriously, for that was her character. She reminded him that the profession of physician, which he hoped to enter, was a serious one, for in years to come the lives and health of thousands of people would depend on his skill. And his ability to acquire the skills of a physician depended upon his acquiring preliminary training at the school of St.-Joseph-in-the-Corner. And in order to be well trained there he had to keep his head clear and free from toys, games, and childish distractions. And his obsession with his birthday was just the kind of distraction he must avoid. She reminded him that he was supposed to be a model his two nephews, Esculapio and Ippocrato, could look up to, not just a little boy no more mature than they.

Tibaldo was overwhelmed by his older sister's speech and would have had difficulty giving her a convincing answer, even though he had practiced debating for the past four years. Fortunately for him there was no opportunity to answer her, for just as she finished her speech there was a loud and urgent knock on the door. The visitor turned out to be the maidservant of Signora Guardabassi, the wife of the third assistant of Governor Domitiani. Anna Maria had begun to take care of her a month previously, when it had become evident that she would soon give birth. Consequently, the sudden arrival of a message from the Guardabassi household was not unexpected. The maidservant, however, was extremely agitated and said, "Signora Anna Maria, come quickly! My mistress is screaming and twisting in terrible pain, and something dreadful will happen if you do not come immediately." Anna Maria always kept a basket ready for emergencies, containing all the medicines, syrups, instruments, and cloths needed at a childbirth. In one minute she fetched the basket, put on her cloak, and was about to depart with the maidservant, when a sudden thought struck her. She said to Tibaldo, "You come with me. You will see what it is like to take care of some one who is suffering, and the lesson will drive childish ideas out of your head. Besides, I may need some assistance."

When they entered Signora Guardabassi's bedroom they found the lady lying on her back in her bed, without a pillow under her head and with the blankets tossed

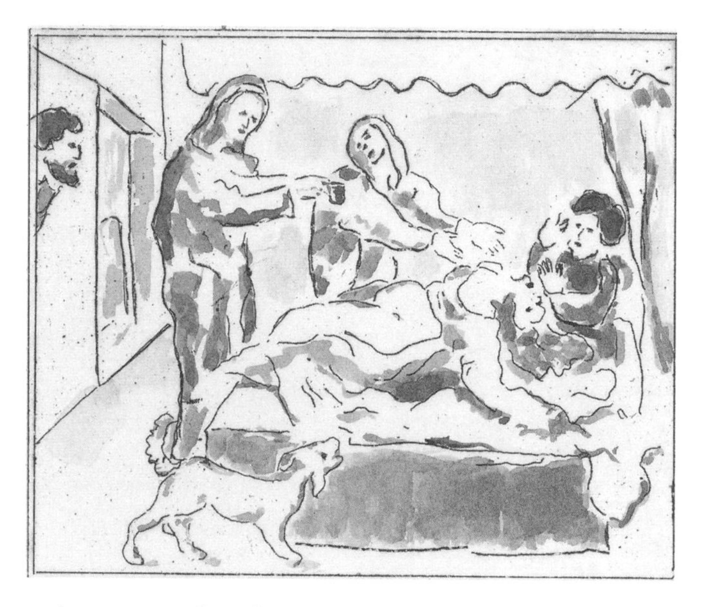

into a heap on the floor. The poor lady was so frantic with pain and fear she had thrown off her bedding. But in spite of the cold, much sweat was flowing down her face and body. She was crying wildly, and when she saw who had arrived she shouted, "O Signora Anna Maria, the pain is so terrible. I shall die, and it is better that I die than suffer." Anna Maria began to talk to the lady soothingly, and at the same time she propped her into a half sitting position, supported by pillows, so the entire weight of the uterus, which contains fluids and her baby, was not pressing upon her spine. She arranged the blankets over Signora Guardabassi to keep her warm, since chill makes the muscles tense and resistant to the normal rhythm of contractions of the uterus. Anna Maria had explained many times to Signor Guardabassi and the maidservant that these simple precautions should be taken as soon as contractions began, but the maidservant was so flustered that she had forgotten all the lessons, and the Signor was so frightened that he wouldn't even enter his wife's bedroom. Anna Maria massaged Signora Guardabassi's back to help her relax, and as she did so she reminded the lady of past instruction: "Relax, dear lady, and do not fight against the contractions. They are nature's way of bringing your baby closer to the opening of your uterus and into the atmosphere. They are not a sign that something is wrong with your body, but just the opposite—they show that your body is doing exactly what it is supposed to do." Then she took from her basket a bottle containing syrup of her own concoction, made of raspberry juice and honey, and gave a cupful of it to Signora Guardabassi. The drink restored some of the energy the lady had lost in her fits of tossing, and its sweetness helped her to relax. Anna Maria ordered the maidservant and Tibaldo to make a fire in the fireplace opposite the bed, a task the boy was glad to perform, since he was stunned by the spectacle of the distraught, sweating, screaming lady, and he needed some distraction.

Anna Maria had not yet examined Signora Guardabassi's uterus, and before she did so she washed her hands carefully in a basin that stood in the bedroom. Then she made her examination and said, "The cervix of the uterus, which is the part at its lower end, has not yet opened very much.

"You are still in the first stage of labor, and the baby won't come for many hours. You must be patient. First babies usually take a long time." By this time Signora

Guardabassi's contractions had stopped, and she was relaxed enough to doze for a while.

After two hours contractions began again, but now more frequently and more intensely. But Signora Guardabassi was so confident of the help of her midwife that she no longer resisted them. Instead, she followed Anna Maria's instructions to press with her muscles to help the contractions open the cervix and bring the baby closer to the aperture. Between contractions she looked at the fire that Tibaldo and the maidservant had made, and its dancing flames had a wonderfully soothing effect on her. Soon Anna Maria was able to see the baby's head and announced the baby was in the normal position, with the head towards the aperture and its face toward the front of the mother's body. She told Signora Guardabassi that this position would ensure the easiest possible birth. By this time the contractions were very strong, and Anna Maria told the mother to stop all pushing efforts, because the natural action of her muscles would bring out the baby. And indeed that is what happened. The baby gave a loud cry when only its head and upper shoulders had emerged, and everyone knew that the birth would be safe. When the baby, now seen to be a girl, completely emerged Anna Maria placed her on her mother's breast, which is what both mother and child wanted by instinct.

A few more things remained to be done, and they were almost as astonishing as what had gone before. Within minutes a partially transparent membrane emerged from Signora Guardabassi's uterus. It contained a mass of tissue, which in fact was the placenta, an organ that had developed in the inner lining of the uterus for the purpose of nourishing the unborn baby and preventing undesirable substances from entering the baby's bloodstream. Between the placenta and the baby's navel stretched the umbilical cord, which was the tube through which the baby was fed while it was still in the uterus. Anna Maria soon cut the umbilical cord near the baby's navel and tied the part attached to the baby into a neat small knot, while the maidservant removed the membrane and the placenta from the bedroom. As Anna Maria washed the mother and her new daughter, Signora Guardabassi said in a low voice, because she was exhausted, "Signora Anna Maria, I am very grateful to you. You are the best friend I ever had, and I shall name my daughter after you."

It was only after Anna Maria had finished her duties as midwife that she took notice of Tibaldo, and she observed that he was pale and trembling. "What is wrong with you?" she asked. Tibaldo was so astonished and shocked by what he had witnessed in the last five hours that he could not answer at first. His sister poured out a cup of her raspberry and honey syrup for him and said, "You need this as much as Signora Guardabassi did when we first arrived." The drink did make him steadier, but he still found it difficult to speak because of the many confused thoughts that crowded into his mind. Some of the thoughts were about the fantastic way that a baby is born, which he had known before in a general way but not in any detail. Some thoughts were about the appearance of a woman, which he found disturbing but could not explain why. Some thoughts were about the terrible agony of Signora Guardabassi when they first arrived, which seemed much worse than the suffering of any of the wounded or sick people whom he had seen his father treating. And some of the thoughts were doubts about himself, especially about his ability to become a physician and heal people in great pain. He finally managed to tell his sister about these doubts. Anna Maria said, "It is a very good thing that you feel that way. Some one who doesn't understand how much people can suffer and how hard it sometimes is to heal them will never become a good physician. Some one who does understand these things is willing to make the great effort to learn the skills of a physician. You will make that effort. You are lucky, too, that you have a sister who is a midwife, and who learned the art of the midwife from both our aunt and our grandmother, because midwives know some things that don't seem to be taught in the medical schools. Did you see how I induced Signora Guardabassi to stop her struggle against natural motions of her muscles? Some pain is unavoidable when people are sick or injured, but the multiplication of pain by fear is wrong. Never forget the lesson you learned today." Anna Maria was exhilarated by her successful attendance on Signora Guardabassi and by the presence of her little brother, who obviously admired her, and she could not resist continuing her lecture: "Did you notice how carefully I washed my hands before examining the lady? And how short I keep my fingernails, to prevent dirt from getting under them, which then may enter the bodies of the mother and the baby? I can't tell why a tiny bit of dirt can do harm to them. Maybe some of the learned professors at the Medical School can give the explanation. All I can say is that a little dirt can do much harm. I know a dozen midwives in Bologna and nearby and a few from farther off, and our aunt and our grandmother knew many more. All three of us observed that the midwives who wash their hands very well and boil their instruments and their cloths have less fever among the mothers and many fewer deaths of infants than the slovenly midwives. It's a fact, and there must be a reason."

Anna Maria might have continued longer, but she saw once more a sad expression on Tibaldo's face. "What is troubling you now?" she demanded. As a matter of fact, it was Tibaldo's obsession with his birthday, which came back to him in spite of the excitement of the day. He confessed to his sister, "That little girl baby is fortunate. She is just born, and it is certain that as long as she lives she will have a birthday each year. And here I am, about to lose my twelfth birthday, which is a very important one." Anna Maria was disgusted with Tibaldo. "You have just witnessed one of the greatest wonders in the world, the birth of a human baby. You have just learned some valuable lessons for your future career as a great physician, and nevertheless you still whine idiotically about a birthday. I thought you were beginning to grow up, but now I wonder whether you ever will."

Both were silent for a while. Anna Maria had said everything that she felt, and Tibaldo felt too embarrassed to say anything more. Then Anna Maria hugged her little brother because she loved him and also had begun to realize that he still was a child, in spite of his excellent performance at school, and therefore it was natural that he should have childish obsessions. She said, "Tibaldo, you are impossibly foolish. You certainly don't deserve any assistance from me with your birthday mania, but I'll be foolishly generous and tell you something that may help you."

What she told him was the astonishing news that Pope Gregory himself planned to visit Bologna in September. Now how should a midwife in Bologna know about the plans of His Holiness in Rome? Well, it is not for nothing that a common name for a midwife was "wise woman." Midwives know things, and they learn things. Here is the explanation. Signor Guardabassi, as we have said, was third assistant to Signor Domitiani the Governor of Bologna. When the Pope decided to visit the city where he had been born, brought up, and educated—because he was sentimental and wanted to visit his native city once more before his death—he of course wrote to the Governor, ordering him to prepare a suitable reception. The Pope did not want news of his journey to be spread around for various reasons, not the least being that there were many bandits infesting the roads of the Papal States. Nevertheless, Signor Domitiani had to tell his assistants, in order to have them plan well in advance to house, protect, entertain, and honor the large cavalcade that would accompany the Pope. Signor Guardabassi was given the great but frightening honor of being the chairman of the planning committee. That was just at the time when Anna Maria began to visit his wife to prepare her for bearing a child. Signor Guardabassi wanted to know whether his expected baby would be old enough and strong enough to be presented to the Pope, since a papal blessing was the greatest gift his child could receive. Consequently, Signor Guardabassi questioned Anna Maria after she had examined his wife to find out when the baby would be born. Would it arrive soon enough before September that it could safely be carried to His Holiness? Now it really was not necessary to tell Anna Maria why he wanted to know the time of

birth of his baby. After all, why should an expectant father not be curious? But of course Signor Guardabassi was very proud of being one of the few people who knew in March about the Pope's plans for a visit in September and even prouder of his position as chairman of the planning committee. Somehow, he could not resist revealing the great news to the midwife, of course making her promise not to tell anyone else.

Anna Maria was properly impressed by Signor Guardabassi's news, and indeed she kept her promise of silence—at least, she kept it until the moment when she revealed it to Tibaldo. She said to him, "It is Pope Gregory who took your birthday from you. No one else but he can give it back. Perhaps you can find a way to persuade him to do so."

Tibaldo was overjoyed at the news that his sister had told him, and he immediately began to think up possible schemes to see the Pope and persuade him. It had been a very full day for Tibaldo.

Pope Gregory's Visit to Bologna

By the beginning of July 1582, it was no longer a secret that Pope Gregory would be coming to Bologna the first week of September. So much had to be done in preparation for his visit that every one was sure to have heard. Bologna is three hundred kilometers to the north of Rome, and the roads through the Apennine Mountains were winding and rough. The voyage could take six days or more, even with the best carriage and horses. It would be a hard trip for an eighty year old man. Nevertheless, Pope Gregory was willing to endure the discomforts of the journey because of his strong feelings for Bologna, where he had been born in 1502.

His name, before being elected pope, was Ugo Buoncompagni, as we mentioned, and it warmed his heart to have throngs of young members of the Buoncompagni family presented to him. When one of his nephews or grand-nephews impressed him as especially talented, he found a position for him in the offices of the Vatican or in the government of the Papal States. In Bologna he was received as a prince who had been absent for a long time from his subjects, because Bologna was one of the great cities of the Papal States and one that was favored due to the Pope's sentimental attachment to it. Bologna was famous throughout Italy for its colorful processions and

ceremonies, and the citizens knew how to devise a celebration of the greatest splendor for Pope Gregory, who was simultaneously their countryman, the ruler of their government, and their supreme religious leader.

At the school of St.-Joseph-in-the-Corner there was a special reason for excitement. Pope Gregory had attended the school, of course long before Master Domenico became principal. He had, in fact, graduated with highest honors in 1517 and then had studied law and theology at the University of Bologna, where he became a professor of law. During his long life he had been the friend of seven principals of St.-Joseph-in-the-Corner. Whenever he visited Bologna he liked to meet a small delegation of students, chosen carefully by the teachers. He loved to hear the long passages of Latin they knew by heart, to tell them about astonishing feats of memorizing that boys were able to perform in his day, to ask them difficult problems of interpretation, and finally to give them his blessing. Pope Gregory was ordinarily a stern and solemn man, but on these occasions he seemed the benevolent father.

There was strong competition among the students to be chosen to meet the Pope. Some of the boys had obvious advantages—they were the oldest, or the ones who had been at the school the longest, or the ones who had won the contests of translating and memorizing and debating, or the ones whose fathers were the most influential, or the ones who had flattered the teachers.

Tibaldo very much wanted the honor of being chosen for the delegation, but he had another motivation, which was his secret. He dreamed that somehow he would have the chance to ask Pope Gregory to rescue his birthday from the hole in the calendar. His obsession was ridiculous, as he was beginning to recognize himself. But that is the way obsessions are. You cannot tell an obsession to disappear just because it is ridiculous. Hence

Tibaldo hoped much more intensely than his schoolmates to be one of the twelve students chosen to meet the Pope.

What were his chances? He was not one of the oldest—he lost on that point. He had not been at the school as long as many others—he lost on that point. He had occasionally won in the contests of translating, memorizing, and debating, for he was very bright indeed—a gain on that point. Tibaldo's father was a poor man with no influence in Bologna—a loss. Did the teachers favor him? Yes, two did: namely Master Demetrios, who prized Tibaldo's ability to think for himself, and Master Vittorio, who noticed his great curiosity about natural phenomena. The other teachers? Just the opposite. The very qualities that Demetrios and Vittorio prized in him the others disliked because they felt that he asked too many questions and was not as respectful as he was supposed to be. One teacher said, "Such boys are dangerous; they think too much." On the point of being favored by the teachers, Tibaldo lost on balance more than he gained. When everything was added up, Tibaldo lost. He was not chosen.

Nevertheless, Tibaldo did not abandon all hope. He had what in later centuries came to be known as a secret weapon. As you recall, Tibaldo's father, Lorenzo Bondi, was Professor Turisanus's assistant at the Medical School of the University of Bologna. Lorenzo felt that his children should know something about how the human body is constructed, whether or not they went on to have medicine as a profession. He used to tell his children that there is no machine or building in the world that is put together more magnificently than the human body. Consequently, he occasionally brought home specimens for them to examine. Sometimes he brought the bones of a hand or a foot or a spine and let the children assemble them with pins and wires, and once he brought a

whole skull. Even more fascinating were bottles of alcohol containing organs, like a heart or a kidney. Strangest of all was a bottle containing a little embryo, only three centimeters long, with a little tail! It was frightening to see at first, but after a while it seemed more wonderful than frightening. Sometimes on a Sunday evening Tibaldo would slip one of the specimens into his school bag to show to his fellow students the next day. Nothing gave Tibaldo more prestige than his occasional exhibits of anatomy. These specimens were his secret weapon.

One of Tibaldo's closest friends was Stefano Costa, also eleven years old, also lively and intelligent, and in fact so similar to Tibaldo in appearance that they were often taken for brothers. In one respect, however, they were very different. Stefano's father, the greatest silk merchant in Bologna, was as rich as Tibaldo's father was poor. He was also favored much more by the teachers and the principal than Tibaldo, and you may suspect that a major reason was his father's wealth. At any rate, Stefano was included in the delegation that would meet Pope Gregory. Along with the other students who were chosen he began to attend special preparation classes, which were coached by Master Domenico himself.

A few days before the Pope's arrival Tibaldo whispered a proposal to Stefano—a proposal for a trade. Stefano was more fascinated than any one else at the school by the anatomical specimens, and Tibaldo was willing to give him one in exchange for his place on the delegation. Tibaldo said that if they switched clothes on the day of the meeting, no one would notice the substitution, because the two boys looked alike. Stefano would still have the honor among his schoolmates of being chosen for the delegation. He would only lose the hour of the meeting itself—and he would gain permanent possession of the specimen.

Stefano was interested, more interested than he would admit, and some hard bargaining began between the two boys. "Which specimen?" asked Stefano. "The bones of a foot," proposed Tibaldo. "That's not worth giving up my meeting with the Pope," replied Stefano. "A jaw with all the teeth in," proposed Tibaldo. "Not good enough." said Stefano. After a while it became clear that Stephano was holding out for the prize specimen of all, the embryo with the little tail. It was Tibaldo's to trade, for his father had given it to him as a special present on his first birthday after Professor Turisanus had sent him off to school. The bottle with the embryo was

Tibaldo's dearest possession. He would have gone without eating for three days in order to keep it. To give it up in order to meet the Pope was a tremendous sacrifice. Poor Tibaldo tried to interest Stefano in alternatives, but nothing worked. Stefano knew exactly what he wanted, and it was just the object that Tibaldo treasured most. Tibaldo was terribly torn. If he had not been obsessed about the loss of his birthday, he would certainly not have agreed to trade it to Stefano. But an obsession takes over the whole mind, and it was victorious even over the beloved specimen. The boys agreed and shook hands, and the bargain was done.

September 7, 1582, was a magnificent day. Pope Gregory celebrated an early mass at the Cathedral of San Pietro and then rode in an open coach drawn by four white horses to the Palazzo Communale. One-quarter of the people of Bologna marched in a procession along with him, and the other three-quarters lined the streets, looked out of windows, climbed trees, and sat on rooftops to watch. There were troops of soldiers on foot, and troops of soldiers on horseback, all dressed in brilliant uniforms of the papal army. Large banners of many colors were carried, some showing the emblems of the Pope, some of the city of Bologna, and some of the Buoncompagni family. There were trumpeters, fife players, and drummers. Each of the guilds of Bologna had a company in the procession: the glass-makers, the builders, the weavers, the blacksmiths, the coppersmiths, the goldsmiths, the butchers, the wine makers, the barrel makers, the bakers, and on and on. Relics of the patron saint of Bologna, San Petronio, were carried under a

canopy held up by ten men. The procession marched by a circuitous route through the great streets with arcades, for which Bologna is famous, and passed the leaning towers of Asinelli and Garisenda. (The citizens of Bologna always complained that Pisa is famous for its one leaning tower whereas their city possessed two, one leaning farther than the tower of Pisa.) Finally, the procession went through the Piazza Maggiore and entered the Piazza Nettuno, in front of the Palazzo Communale, which was the center of government of the city of

Bologna. A statue of Pope Gregory had been erected above the gateway to the Palazzo Communale only two years earlier, and this was the first occasion for the Pope to admire it in its permanent place.

In front of the gateway a great wooden platform had been erected, in the center of which was a wooden throne covered with gold and shaded by ostrich feathers. For two hours one delegation after another reverently presented itself to the Pope, bringing gifts, making petitions, and receiving blessings. The guild of glass-blowers brought him a beautiful goblet, the guild of builders a model of the Church of San Petronio, the weavers an embroidered red cape, and so on from all the guilds. It was strenuous for a man of eighty to sit so long and meet so many delegations. But Pope Gregory felt that he was among his own people, and he loved the ceremony in spite of the strain.

Finally the time came for the delegation from the school of St.-Joseph-in-the-Corner. Pope Gregory looked forward most to its appearance, because of fondness for his old school. Another person who eagerly looked forward to seeing this delegation was Signor Costa, the father of Stefano, who stood at the front of the crowd facing the the platform.

He was gratified that his son had been chosen as one of the twelve members of the delegation, and he had hinted to Master Domenico that he would be pleased if his son were there. For the occasion he had hired the most expert tailor in Bologna to make proper ceremonial clothing for his son: breeches, a jacket, and a tunic, all of embroidered blue silk, and a wide-brimmed velvet hat. He swelled with pride when the delegation emerged from the gate of the Palazzo Communale and mounted the staircase of the platform, for no other student was dressed so richly and tastefully. But when the students in the delegation removed their hats in reverence to the Pope, Signor Costa was astonished to see not his son Stefano but young Tibaldo Bondi, similar indeed but certainly distinguishable by a parent's eye. He whispered to his wife standing at his side. "That is the Bondi scamp, disguised as Stefano! And where is Stefano? This is an outrage and a crime, and I will expose the scandal immediately!" Signora Costa held him firmly by the arm and whispered back, "My lord, do you know what Signor Domitiani would do to you if you interrupt the ceremony of the Pope?" In spite of his fury Signor Costa controlled himself and made no disturbance. He did not answer his wife, which is as close as he could ever come to admitting to her that she was right. He bitterly

watched the events that followed, resolved to punish young Bondi at the right time, and Stefano as well if he was partly to blame. Tibaldo was unaware how perilously his prospects tottered at that moment. Had Signor Costa exposed him as an imposter, he would surely have been expelled from the school, fallen out of favor with Professor Turisanus, lost his chance to become a medical doctor, and made his parents and himself miserable for the rest of their lives.

Instead, the audience with the Pope proceeded as planned. Several of the students recited passages from Cicero, Virgil, St. Augustine, and St. Thomas Aquinas. At first Pope Gregory listened with approval, but after a particularly long recitation his attention began to wander. His head appeared to nod a little and his eyes drooped—both bad signs. But then he roused himself and suddenly asked, "Can any one recite anything that We have written?" (Popes, like kings, are permitted to say "We" when they mean "I".) Almost all the students were taken by surprise, because Master Domenico had not thought to prepare them for this question. But Tibaldo volunteered, saying bravely, "I can, Your Holiness."

He began to recite: *"Inter gravissimas pastoralis officii nostri curas, ea postrema non est, ut quae ab sacro Tridentino Concilio Sedi Apostolicae reservata sunt, illa ad finem optatum, Deo adiutore, perducantur...."* and continued to say twenty lines of the Pope's proclamation on the reform of the calendar. After all, Tibaldo had studied that document very carefully, since it affected a vital part of his life. The

Pope's eyes glowed with pride, both for the beautiful Latin prose he had written and for his old school, which had produced a student who displayed such remarkable knowledge.

Pope Gregory then asked Tibaldo, "Young man, there has been much dispute about our new calendar. What is your opinion?" It was a glorious opportunity for Tibaldo, and he knew that he must proceed carefully and skillfully. That was particularly hard to do, for he was trembling as he faced the stern, powerful old man who was both head of his Church and ruler of his City. Then he remembered Master Demetrios's admonition, "Never be afraid!" and he began to speak, humbly and softly but steadily. "Your Holiness, the new calendar is a great improvement. The old calendar was contrary to nature. It made the legal year different in length from the natural year, and the seasons and the holidays were becoming more and more confused." Pope Gregory gave a gesture of agreement, for that was exactly why he had changed the calendar.

Tibaldo continued, "But, Your Holiness, there is a hole in the calendar. And there are days that have fallen through the hole." The eyes of Pope Gregory narrowed, and the smile disappeared from his face. He did not appreciate having any one use the word "but" to him. "Explain yourself, young man," he said.

"Your Holiness, every anniversary and every birthday between October 5 and October 15 will be lost in the year 1582, because those dates will be skipped, and...."

Tibaldo was not allowed to finish his sentence. The

Pope asked sharply, "Is a birthday an important thing when we are restoring order to the year? Small things must be set aside when great things are accomplished."

The crucial moment had come, and Tibaldo almost faltered. But he continued, "Your Holiness, almost everybody is unimportant in a great city. One day a year, however, family and friends pay special attention to a person, who then feels important for that one day. It is impor-

tant that every one should occasionally feel important." The years of debating at school had indeed sharpened Tibaldo's wits.

The Pope began once more to be less stern with the boy, and a slight look of amusement came into his face, but he raised another objection. "Young man, a birthday is a silly, frivolous thing of the world. If a person has to be important one day a year, let it be on his name day, which is a sacred occasion." The name day, it should be explained, is the feast day of the saint after whom the person was named.

Tibaldo was caught by surprise, but in a second he thought of an answer. "Your Holiness, think of the poor people who will lose their name days due to the hole in the calendar, because the feasts of their saints fall between October 5 and October 15, and those days will not exist in 1582. Think of the poor people named after Saint Apollinaris, Saint Bruno, Saint Justina, Saint Marcellus, Saint Denis, Saint Francis Borgia, Saint Andronicus, Saint Ethelburga, Saint Faustus, and Saint Callistus. Do you not feel sorry for them?" And then one more thought came into his head, "And will not Saint Apollinaris and Saint Bruno and Saint Justina and Saint Marcellus themselves and all the others feel sad, even though they are in heaven, because their feast days have been taken away from them?"

Pope Gregory began to laugh—a very rare thing for him, the first time in fact in eight years. He said, "Young man, you have a good mind and a quick wit. The school of St. Joseph-in-the-Corner should be very proud of you. We predict that you will have a great future." And then he turned to the secretary who always stood at his side and ordered him to write the following addition to his proclamation of the change of the calendar:

"Second only to our concern, as Apostolic Father, for the eternal salvation of the faithful is our care for their lawful and proper earthly well being. Anniversaries, celebrations, and civic festivals being the occasions of proper rejoicing and refreshment of the spirit, it is contrary to our intention that any of these be omitted in consequence of the reform of the calendar which we proclaimed from the Holy Seat on February 24 of the year 1582. We therefore command that all anniversaries etc., customarily occurring between the fifth day of October and the fifteenth day of October, which days are omitted in the new calendar for the year 1582, be celebrated on their customary days according to the old calendar for the year 1582 alone. All anniversaries, etc., customarily occurring on the fifteenth of October or thereafter should be celebrated according to the new calendar."

Tibaldo bowed, thanked the Pope humbly, and received a blessing. Then the delegation from the school was dismissed and the ceremony ended. The aged Pope was very tired and needed to rest, but he continued to be amused as he left the platform and returned to his chamber.

Tibaldo, of course, was triumphant. He had accomplished exactly what he had intended to do and had done so with great courage and cleverness. His obsession was now satisfied. He would have his twelfth birthday after

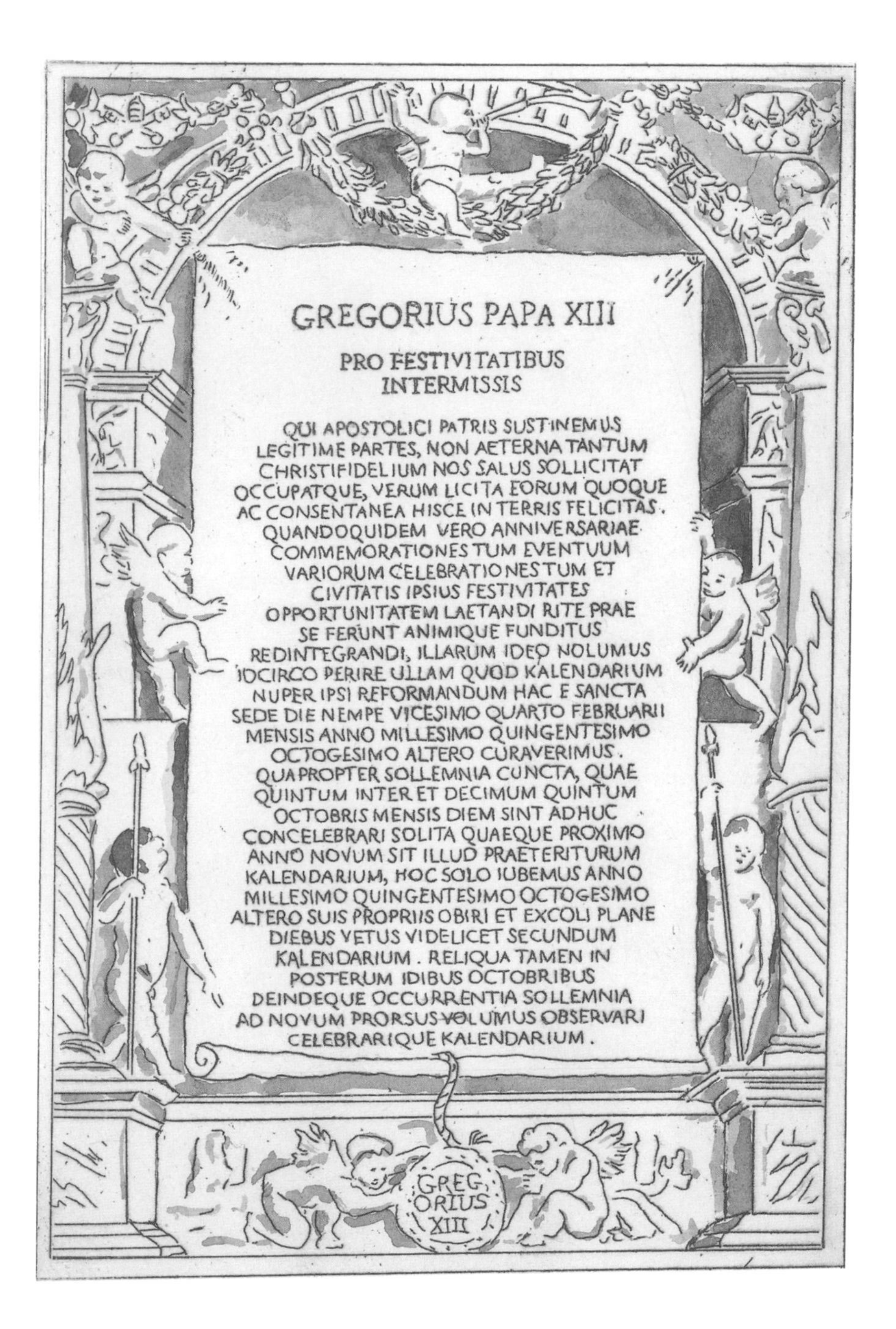
GREGORIUS PAPA XIII
PRO FESTIVITATIBUS
INTERMISSIS
QUI APOSTOLICI PATRIS SUSTINEMUS
LEGITIME PARTES, NON AETERNA TANTUM
CHRISTIFIDELIUM NOS SALUS SOLLICITAT
OCCUPATQUE, VERUM LICITA EORUM QUOQUE
AC CONSENTANEA HISCE IN TERRIS FELICITAS.
QUANDOQUIDEM VERO ANNIVERSARIAE
COMMEMORATIONES TUM EVENTUUM
VARIORUM CELEBRATIONES TUM ET
CIVITATIS IPSIUS FESTIVITATES
OPPORTUNITATEM LAETANDI RITE PRAE
SE FERUNT ANIMIQUE FUNDITUS
REDINTEGRANDI, ILLARUM IDEO NOLUMUS
IDCIRCO PERIRE ULLAM QUOD KALENDARIUM
NUPER IPSI REFORMANDUM HAC E SANCTA
SEDE DIE NEMPE VICESIMO QUARTO FEBRUARII
MENSIS ANNO MILLESIMO QUINGENTESIMO
OCTOGESIMO ALTERO CURAVERIMUS.
QUAPROPTER SOLLEMNIA CUNCTA, QUAE
QUINTUM INTER ET DECIMUM QUINTUM
OCTOBRIS MENSIS DIEM SINT ADHUC
CONCELEBRARI SOLITA QUAEQUE PROXIMO
ANNO NOVUM SIT ILLUD PRAETERITURUM
KALENDARIUM, HOC SOLO IUBEMUS ANNO
MILLESIMO QUINGENTESIMO OCTOGESIMO
ALTERO SUIS PROPRIIS OBIRI ET EXCOLI PLANE
DIEBUS VETUS VIDELICET SECUNDUM
KALENDARIUM. RELIQUA TAMEN IN
POSTERUM IDIBUS OCTOBRIBUS
DEINDEQUE OCCURRENTIA SOLLEMNIA
AD NOVUM PRORSUS VOLUMUS OBSERVARI
CELEBRARIQUE KALENDARIUM.
GREG
ORIUS
XIII

all. It would be celebrated on October 10 according to the Julian calendar, even though that day would be labeled October 20 according to the Gregorian calendar The crucial thing was that it would take place and that Pope Gregory had made it official.

There was a flaw in his triumph, however. Signor Costa was furious that Tibaldo had exchanged clothes with his son Stefano and had slipped into the delegation. Even before the crowd dispersed he found the principal of St.-Joseph-in-the-Corner and demanded that Tibaldo Bondi be severely punished for his slyness and deception. Master Domenico summoned Tibaldo to his office and scolded him ferociously. He was considering what pun-

ishment to give the boy. Should Tibaldo be expelled from the school? Should he be whipped twenty times in front of all the students? Should he be placed on a diet of dry bread and water for a month? Fortunately for Tibaldo, an assistant knocked at the principal's door while he was trying to decide on a punishment.

A letter had just arrived from Pope Gregory congratulating Master Domenico on the excellence of his students and particularly asking the name of the bright boy with whom he had discussed the calendar. Tibaldo was saved. Master Domenico could hardly punish him at just the moment when the Pope had sent him special congratulations.

Consequently, he swallowed his anger, forgave Tibaldo, and reluctantly congratulated him. Then Master Domencio had the unpleasant task of informing Signor Costa that no action could be taken against Tibaldo because the Pope had personally intervened to save him. Signor Costa had to content himself with scolding Stefano, who did not have papal protection.

But Tibaldo still had to face his father, Lorenzo Bondi, who was very upset that his son had taken risks that could have resulted in his being expelled from school and deprived of the possibility of a great career. "How could you?" he kept repeating. "How could you?" But Tibaldo's mother, who was always the peace-maker in the family, calmed her husband: "Look what great honor our son has brought upon himself, upon you, and upon our whole family by his brilliant recitation before the Pope. Every one in Bologna admires him. Do not always worry about what might have happened. Look at the good thing that did happen." So Tibaldo's father calmed down, and there was peace and rejoicing in the Bondi household.

What Happened Afterwards

On the following day, September 8, Pope Gregory's addition to his proclamation changing the calendar was printed in Bologna, and copies were sent all over Italy and to other countries. He provided a safety net for all those birthdays, anniversaries, and festivals that were in danger of falling through the hole in the calendar. Hence October of 1582 was a strange month for celebrations. On the day that was October 5 according to the old calendar but October 15 according to the new, festivals of both days were celebrated. On the following

day, festivals of both October 6 and October 16 were celebrated. And so on to the tenth day, when the festivals dated both October 14 and October 24 were celebrated. During this period each day performed a double duty. The bakeries and candy shops and wine shops and sausage shops and toy shops did much more business than usual. Consequently, the period in which ten days were omitted in the new calendar turned out to be a time of rejoicing instead of sadness. The Pope's additional proclamation

in Bologna was exactly what was needed to rescue the endangered celebrations. And Tibaldo was responsible for their rescue. He was a hero.

Tibaldo was at home on his birthday (October 10 according to the old calendar, even though it was October 20 according to the new one). His mother prepared an even more special and delicious dinner than the ones she usually prepared for birthdays. As a present, his

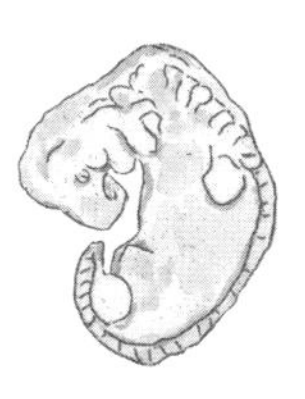

father brought him a new specimen in a bottle of alcohol from the Medical School. You can guess what it was. Tibaldo's happiness was complete.

There was general contentment with the new calendar of Pope Gregory in all Catholic countries, where he had authority to enforce it. In the Protestant countries there was resistance to changing the calendar, precisely because it was a Commission appointed by a Pope who had suggested the changes and a Pope who had proclaimed them. But eventually the flaws of the old Julian calendar were perceived all over the world. In Great Britain and the British colonies in America, for example, the change was made in 1752, by which time eleven days had to be dropped. In Russia it was not until 1919 that the change was made, and then thirteen days had to be dropped. As a result we have the peculiarity that the Bolshevik Revolution in the fall of 1917 is sometimes called the "October Revolution" (by the Julian calendar) and sometimes the "November Revolution" (by the Gregorian calendar).

Nothing ever happened again to Tibaldo as unusual as his interview with Pope Gregory, but still he had a very interesting life. Five years after his great adventure he graduated from the school of St.-Joseph-in-the-Corner and with the help of Professor Turisanus entered the Medical School of the University of Bologna. Years of hard studying followed, most of it not at all inspiring, because he had to listen to lectures on superstitious medical theories and astrology and the supposedly health-giving properties of hundreds of disgusting substances. But Tibaldo was tenacious. He knew that these studies were for the most part obstacles that had to be overcome on the way to becoming a medical doctor, and he tried, by thinking for himself, to pick out from all that he was taught the things most likely to be helpful for treating patients in years to come. He kept his eyes open and learned the skills of his father, and he kept in mind the advice of his sister Anna Maria to observe which treatments work and which do not. In this way he learned more from experience than from books and lectures. He asked many questions and had doubts about the authorities. But he was careful not to say in public that the authorities often did not know what they were talking about. Had he done so, he might have been expelled from the Medical School and lost his chance to become a physician. It was not easy to be a young man who thought for himself in a period of superstition and authority.

After Tibaldo received his medical degree he was chosen as Professor Turisanus's successor. Tibaldo was grateful to the old Professor for providing his education, despite doubts about most of his lectures, and he allowed Professor Turisanus to regard him as a substitute son. The people from whom Tibaldo learned the most were his father Lorenzo, his sister Anna Maria, and two teachers, Masters Demetrios and Vittorio, at his old school.

Once Tibaldo (or Professor Bondi, as he should perhaps be called) had reached this position he was free to try out medical ideas he had partly learned from others and partly had developed himself over many years. He believed very firmly that the human body has more ability to cure its own disturbances than most physicians would grant. It was therefore most important for a physician to support the natural healing powers of the body and not to weaken them by unnecessary intrusions. He liked to quote the advice the great Hippocrates gave to physicians nearly two thousand years earlier: "Above all, do no harm."

Consequently, he would not prescribe that patients be bled, even when they had high fevers, by either cutting their veins or applying leeches, because he had observed that bleeding weakened patients. Likewise, he would not give emetics to induce vomiting unless he had good evidence that the patient had swallowed something poisonous, because prolonged vomiting drew water from the body and exhausted the patient, nor would he prescribe enemas unless there was evidence of blockage of the bowels. He himself never applied hot irons to wounds in the hope of cleaning them, and he tried to persuade other physicians to use fresh bandages instead. He made patients as comfortable as possible with pillows, warm blankets, and good nourishment; and he encouraged them to be cheerful and optimistic. Some of the other physicians in Bologna were envious of his success and said, "Dr. Bondi does nothing, and then takes the credit when his patients recover by themselves."

When Tibaldo heard this gossip, he said, "That's just as it should be. The surgeon Ambroise Paré was right when he said, 'I dress the wounds, and God heals them.'"

When Tibaldo taught his students, he preferred not to use the lecture rooms of the Medical School but took his class instead to the hospital of Bologna to see patients in various stages of sickness. The medical students could see for themselves the symptoms of various diseaseas, the effects of different treatments, and the normal rate of recovery from a disease. In the hospital room Tibaldo could point out to his students how many precautions were needed to prevent pollution from entering the bodies of the patients: that bandages and instruments had to be boiled, water and food had to be shielded from indirect contact with the excretions of other patients, and the hands of physicians and attendants had to be thoroughly washed. He told his students, "Sometimes nothing can be done to save a patient from a disease, but at least we can prevent the disease from spreading to others."

Tibaldo often brooded on the remark of his sister Anna Maria that maybe the learned professors at the university could explain why dirt was dangerous to a new born baby and its mother. He had never heard a convincing explanation while he was a medical student, but he read with great interest the theory of Doctor Girolamo Fracastoro of Verona that many diseases are caused by the entrance of tiny particles into the patient's body, some of which are alive and capable of multiplying.

When Tibaldo learned about the invention of an instrument combining lenses, called the microscope, which could produce visible images of tiny things, he had one made for himself by a lens grinder so he could search for Doctor Fracastoro's tiny organisms. To his great disappointment he never saw anything more than little specks of dust which did not behave at all like organisms. In 1680, however, eighteen years after Tibaldo's death, the Dutch microscopist Anton van Leeuwenhoek described bacteria that he observed with a microscope much more powerful than the one used by Tibaldo.

The grown up Tibaldo had obsessions, just as the child Tibaldo did, and one of them was the microscope. This was a healthy obsession, and it spread among his colleagues at the University of Bologna. One fine result was that near the end of Tibaldo's life he was able to look at an inspiring demonstration of blood flowing in the living capillaries of a frog's webbed foot, discovered with a microscope by a young professor in Bologna named Marcello Malpighi.

Some other important things happened in Tibaldo's life. In the year 1607 a comet was sighted that became brighter and brighter as the weeks passed, until it was the brightest object in the night sky other than the moon. And after a while, it developed a glowing tail that was longer than the diameter of the moon. Tibaldo had not lost interest in astronomy since the time of his classes with Master Vittorio Rhaeticus, and he often went out at night to observe this comet. One day he noticed an announcement of a lecture on the comet by Sebastiano Tramontano, an astronomer at the University of Bologna, who promised to demonstrate out of doors how the position of the comet in the sky is measured precisely with an instrument called an astrolabe, so that the path of the comet among the stars over a period of many weeks could be plotted. Tibaldo came at the appointed hour to a meadow outside one of the city gates and joined a group of people curious about the nature of the comet and the procedures for observing it. The lecture was very instructive. He was especially interested to hear the theory that the tail of the comet was caused by light from the sun pressing the smaller particles of the comet in a direction away from the sun.

After a while Tibaldo noticed that the assistant who was operating the astrolabe and writing numbers in a notebook was a young woman. Tibaldo thought to himself that as fascinating as the comet was, the assistant was the most beautiful astronomical phenomenon he had ever seen. He later found out that her name was Elisabetta Tramontano, the daughter and best student of the astronomer. For several nights following, Tibaldo came to demonstrations to see the changes in the location of the comet and in the length of its tail, and for other reasons as well. Tibaldo was very much drawn to the young woman, and after a while she noticed him and was drawn to him as well. A few months later they were married.

Tibaldo and Elisabetta continued to look at the objects in the heavens even after they were married and had children. A great occasion was their fourth wedding anniversary in 1611, when Tibaldo presented Elisabetta with a telescope, constructed by a lens maker of Bologna after the design of Galileo's pioneering astronomical telescope of 1610. You see, Tibaldo had never ceased to be fond of birthdays and anniversaries. A small closet was constructed on the roof of the Medical School to store the telescope, so that Tibaldo and Elisabetta could take it out to use at a high elevation, where they had a clear view of most of the sky. They saw wonderful things, like four moons of the planet Jupiter, the phases of the planet Venus, the mountains and valleys of the moon, and a strange band about the equator of the planet Saturn that extended far beyond the planet's surface. Many years later they liked to say, "Our real wedding rings were the rings of Saturn."

Like Professor Turisanus, Tibaldo and Elisabetta had only girls as children. They did not grieve over this fact, as he had done, but disregarded the customs of the time and brought all three of the girls up to have professions.

The oldest was named Anna Maria, after her Tibaldo's oldest sister, and she became her aunt's apprentice and later an expert midwife.

The second daughter, Aureliana, shared her mother's passion for astronomy and her marvellous aptitude for mathematical calculations, and she was trained as an astronomer.

The third daughter, named Teresa after Tibaldo's mother, was the first woman ever to receive the degree of doctor of medicine at the University of Bologna.

In many ways it was the beginning of a new age. The Bondi household continued, however, the great tradition of celebrating birthdays. None of the three daughters ever missed the celebration of her birthday, when she was queen for the day in her home.

And on those occasions Tibaldo's daughters were sometimes able to persuade their father to recollect how he almost lost his twelfth birthday but managed to recover it.

What Happened Afterwards

Tibaldo and Elisabetta lived a long time, but inevitably sickness and frailty afflicted them. Even so, until nearly the end of their lives they would climb slowly up the winding steps of the Medical School, take the telescope out of its closet, and observe the objects of the sky. They used to say,

Looking at the great world takes one's thoughts away from the troubles of the little world.

More and Better Astronomy

Two Lectures by Master Vittorio

Master Vittorio's grandfather, Georg Joachim Rhaeticus, assisted his teacher, Copernicus, in writing the great book *On the Revolutions of the Heavenly Spheres*. Vittorio had inherited his grandfather's copy of this book and read it many times. He was not convinced by all the details of Copernicus's book and thought very hard about ways it might be improved. Nevertheless, he was fully convinced of Copernicus's fundamental propositions that the earth rotates on its axis once a day and revolves in an orbit about the sun once a year. The following two lectures, one about the seasons and the other about the appearance of stars to observers at different locations on the earth, were based on these two propositions, even though Master Vittorio was careful not to mention Copernicus's name at the school of St.-Joseph-in-the-Corner. Later we shall see how Master

Vittorio answered objections that various astronomers had raised against Copernicus's propositions. Some of his answers were conjectures which could not be proved by using the evidence available during his lifetime. However, there was much insight in his guesses, and on the whole the later discoveries of astronomy showed that his conjectures were nearly correct.

Once a year, almost as regularly as the winter solstice, Master Vittorio gave a lecture explaining the cycle of the seasons. He liked to give that lecture around the time of the winter solstice when the cold weather and the long hours of darkness made his students wonder about the cause of seasons. Why should the earth's position in one part of its orbit about the sun cause spring in the northern hemisphere, another position cause summer, another cause autumn, and yet another cause winter? Some students suggested that the earth is not the same distance from the sun throughout the year, and that the sun warms the earth more when the earth is closer to the sun than when it is farther away. But this suggestion did not work because it was already known from travelers' reports in the sixteenth century that the months of summer in the northern hemisphere are the months of winter in the southern hemisphere, and vice versa. If winter in the northern hemisphere were due to increased distance of the earth from the sun, how could one account for summer in southern Africa and in the tip of South America? Master Vittorio gave a much better explanation. The axis of rotation of the earth, he explained, is not perpendicular to the plane in which the earth revolves about the sun. The axis of rotation is tilted at an angle of about twenty-three and one-half degrees to the plane of revolution. Furthermore, as shown in Diagram 1, the direction of this axis relative to the stars is almost unchanged throughout the year. (Master Vittorio did not have a good explanation for the nearly constant direction of the axis of rotation and he was not satisfied with the rather cumbersome explanation that Copernicus proposed, but both of them were correct about the fact that the direction is constant.) In Diagram 1, position A is the position of the earth around December 21. When the earth is in this position, its north pole is tilted away from the sun and its south pole is tilted toward the sun. Hence no sunlight reaches the north pole throughout the twenty-four hours of the day, while sunlight reaches the south pole throughout those twenty-four hours. In fact, we can say more: through the whole twenty-four hours no sunlight will reach any point within twenty-three and a half degrees of the north pole. (The circle that bounds this region—the circle that lies exactly twenty-three and a half degrees south of the pole—is called the arctic circle.) On the other hand, every point within twenty-three and a half degrees of the south pole (every point within the antarctic circle) will be illuminated for the entire twenty-four hours. What can one say about points on the earth's surface in the northern hemisphere but south of the arctic circle? Well, each of these will be illuminated less than twelve hours, and the nearer the point is to the

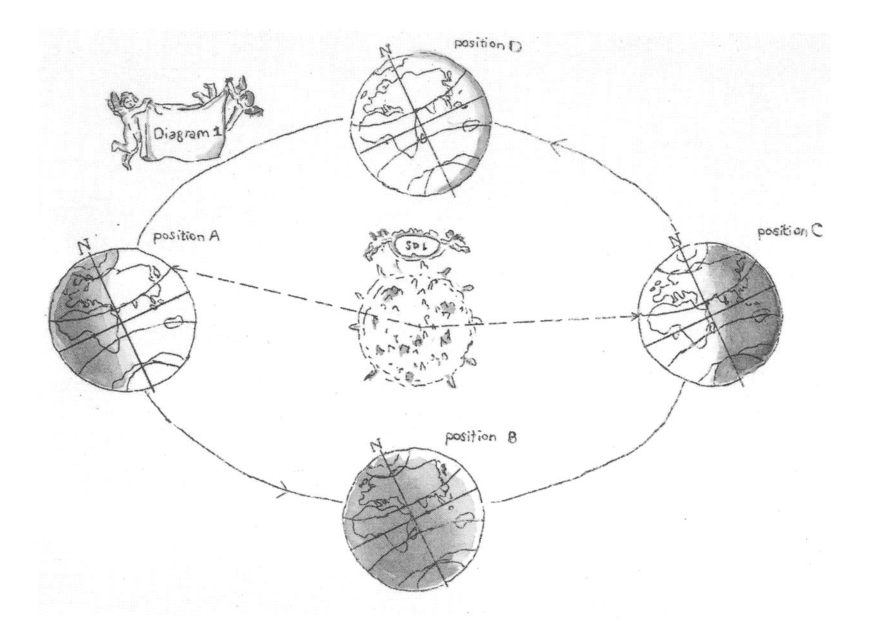

arctic circle the shorter will be the period of daylight and the longer the period of night. In the southern hemisphere just the opposite is true, with longer daylight than night throughout the hemisphere, and more daylight as one goes farther south until one reaching the antartic circle, where at this season the sun never sets. Position A is therefore a winter day in the northern hemisphere and a summer day in the southern hemisphere. The earth is at position A on the day of shortest daylight in the northern hemisphere, which is called the *winter solstice,* while this is the day of longest daylight in the southern hemisphere, hence the *summer solstice* of that hemisphere.

The earth is at position C around June 21. Then the north pole of the earth is tilted towards the sun and the south pole is tilted away. Using the same reasoning we have just been through we find that when the earth is in this position in its orbit there is the summer solstice in the northern hemisphere and the winter solstice in the southern hemisphere. Master Vittorio emphasized that the direction of the axis of the earth's rotation relative to the stars is the same whether the earth is at position A or at position C. The constant direction of this axis and the difference in position of the earth relative to the sun at different times of the year account for the difference in the illumination of the northern and southern hemispheres. In December the southern hemisphere is warmed more by the sun than the northern hemisphere not only because the former has more hours of daylight but because it receives illumination that is more direct and less slanting. Similarly, the northern hemisphere is warmed more than the southern hemisphere in June.

Positions B and D are halfway between A and C. In these positions neither the north pole nor the south pole is tilted towards the sun, and daylight is not favored over night in either the northern or the southern hemispheres. These are the positions of the vernal and autumnal equinoxes, when the periods of daylight and of night are equal almost everywhere on earth. Very near the north pole and the south pole, however, it would not be correct to say that at the equinoxes there are twelve hours of daylight and twelve hours of darkness. What actually happens is very strange indeed. An observer at one of the poles would see half of the disc of the sun going around the entire circle of the horizon during the period of twenty-four hours, while the lower half of the disc is invisible below the horizon. In other words, at every moment of the day it would be half daylight and half night. Of course, in the sixteenth century no one had ever seen this phenomenon, since explorers did not succeed in reaching the north and south poles until early in the twentieth century but good astronomers like Master Vittorio knew what would be seen by any one who reached the north or south pole on a clear day and had the leisure to observe the horizon for twenty-four hours.

Another of Master Vittorio's lectures explained the appearance of the stars in the sky at different latitudes and at different times of the year. Imagine the axis of rotation of the earth to be extended to the sky. The point in the sky above the earth's north pole is called the *celestial north pole,* and the point above the earth's south pole is called the *celestial south pole*. There is a star, called the *North Star* or *Polaris,* very close to the celestial north pole, and it barely moves during the course of a night. If one imagines a plane through the equator of the earth extended indefinitely outward, it will intersect the sky in a circle. This is called the *celestial equator.* An observer standing at the north pole will see all the stars in the sky to the north of the celestial equator (provided, of course, that the sun is not shining and there are no clouds in the sky). The effect of the earth's daily rotation is to make each visible star appear to travel once a day in a circular path in the sky. This circular path is not tilted with respect to the celestial equator, and therefore the star stays at the same height above the horizon throughout its daily course, never rising or setting. This phenomenon is depicted in Diagram 2a. As every one should know, there are nearly six months of daylight at the north pole, when the sun is above the horizon, and nearly six months of night when the sun is below the horizon. Twice a year there is a transition lasting about two and a half days when part of the sun is above the horizon and part below it. The stars in their circular paths are seen only during the period of night. An observer at the south pole will have a similar view of the stars, except that the visible stars are those to the south of the celestial equator, as indicated in Diagram 2b.

An observer on the earth's equator sees the celestial north pole at a point due north on the horizon and sees the celestial south pole at a point due south on the horizon. Since the earth rotates about its axis, the sky seems to turn about an axis stretching between two opposite points on the horizon, and every star is above the horizon for half the day and below the horizon for half the day, as indicated in Diagram 2c. The path of a star through the sky is a semicircle perpendicular to the plane of the horizon, but the star will be visible only if it is above the horizon and the sun is not shining.

At points on the earth's surface in between the equator and the poles the situation is complicated, as shown

in Diagram 2d. For an observer in the northern hemisphere stars close enough to Polaris never sink beneath the horizon but make circles in the sky with Polaris (or more accurately, the celestial north pole) as the center.

Stars close enough to the celestial south pole never rise above the horizon of an observer in the northern hemisphere, and other stars rise above the horizon during part of the day and sink below the horizon the rest of the day. Of course, even the stars above the horizon are visible only in the absence of sunlight. The exact meaning of the phrase "close enough" depends on the latitude—the nearer the observer is to the north pole the larger is the region around Polaris in which stars never set. The apparent paths of the visible stars are either complete circles or arcs of circles—larger arcs for those nearer to Polaris than for those farther from it— and for an observer at a fixed point on the earth's surface the circles are all tilted at the same angle relative to the horizon. The same description of the apparent paths of the stars holds for an observer in the southern hemisphere, but with the celestial south pole substituted for the celestial north pole.

At Bologna, which is almost exactly halfway between the equator and the north pole, the Big Dipper is a constellation that never sinks below the horizon but instead makes a circuit around Polaris. Its circuit can be used as a time-keeper. Suppose that we take the initial time to be the moment when the line from the two outermost stars of the handle of the Big Dipper is oriented due west; then one complete rotation of the earth will have occurred when that line of stars are next in the due west orientation. This time interval is called the *sidereal day* because the Latin word for constellation is *sidus*. It is a remarkable fact that the sidereal day is not equal in length to the solar day. A solar day is the interval between two successive noons, noon being the time when the sun achieves its highest point in its daily course throughout the sky. The solar day is about four minutes longer than the sidereal day. Master Vittorio was able to use Copernicus's theory to give a simple explanation for this difference. The earth's rotation makes the pattern of the stars appear to rotate daily from east to west. The revolution of the earth about the sun causes an eastward apparent circuit of the sun through the ecliptic in one year and since there are 360 degrees in a circle the sun appears to move eastward among the stars a little less than one degree of arc per day. (For this calculation it makes little difference whether the day is taken to be the solar or the sidereal day.) Consequently, when one sidereal day has elapsed from a specified noon, the sun is about one degree of arc east of the posiition of its next noon. But one degree of arc is one three hundred and sixtieth of a complete circle, and four minutes is one three hundred and sixtieth of twenty-four hours. Consequently, four minutes more than the sidereal day are required for the sun to complete the cycle from noon to noon.

Master Vittorio's Answers to Criticisms of Copernicus

Although Master Vittorio revered Copernicus above all other astronomers, he was aware that not all the objections to Copernican theory were merely thoughtless repetitions of ancient scientific opinions or pious quotations of biblical statements. He brooded continually on these objections, trying to find answers to them if he could and considering modifications of Copernican theory if the objections could not be answered.

A serious objection was posed a generation after Copernicus's death by Tycho Brahe, a Danish astronomer who was the most careful, precise, and thorough gatherer of celestial data in all human history up to his time. Tycho argued that Copernicus's theory of the motion of the earth around the sun has consequences which conflict with the astronomcal evidence. The evidence was the failure to see changes of certain angles in the heavens, which were anticipated if the earth moved around the sun and if other assumptions were true. Tycho paid particular attention to the position of Polaris at three different seasons: the autumnal equinox, the winter solstice, and the vernal equinox. If Polaris were located exactly at the north celestial pole, then its position in the heavens would be unchanged throughout the night since the sphere of the heavens appears to rotate on the axis from the south to the north celestial pole. But since Polaris is slightly displaced from the north celestial pole, it appears in the course of a night to go in a small closed path about the pole and hence its altitude (angle above the horizon) varies through the night. On the assumption that the stars occupy a spherical shell not much farther from the earth than Saturn, Tycho was able to calculate that the maximum altitude would change by several minutes between any two of these seasons (a minute being one-sixtieth of a degree), and likewise for the minimum altitude. The observed maximum altitudes on the nights of the autumnal equinox and the winter solstice were both fifty-eight degrees and fifty-one minutes; and the observed minimum altitudes on the nights of the winter solstice and the vernal equinox were both fifty-two degrees and fifty-nine and one half minutes. No changes were seen either of the maximum altitude or the minimum altitude. Tycho was unwilling to admit that the stars might be vastly farther from the earth than the planet Saturn, because if they were, then there would be an immense amount of vacant and hence wasted space, contrary to the wisdom of God in creating the universe. He therefore concluded that Copernicus was wrong in supposing that the earth traveled in an orbit about the sun.

Now Copernicus already had an answer to this objection even before Tycho's birth. Indeed, the answer had already been given by the Greek astronomer Aristarchus around 300 B.C. The answer is that the stars lie far beyond Saturn, so far, in fact, that the change of angle between two stars due to the motion of the earth could not be seen with the instruments available to Aristarchus, or to Copernicus, or even Tycho. This answer was plausible to Master Vittorio, but it was not definitive because no one had been able to prove at his time that the stars are immensely far from the earth. How joyful Copernicus and Master Vittorio would have been had they known about discoveries made independently by three astronomers in the year 1838: F.W. Bessel observing a star in the constellation Cygnus; F. von Struve observing the star Vega; and T. Henderson observing the star Alpha Centauri in the southern hemisphere. They all made the assumption that most of the dim stars are much farther from the earth than the bright stars, which discarded Tycho's idea (shared by Copernicus, Aristarchus, and most of the early astronomers) that all the stars are confined to one spherical shell beyond the planets; but that step had already been taken by the English Copernican Thomas Digges in 1576.

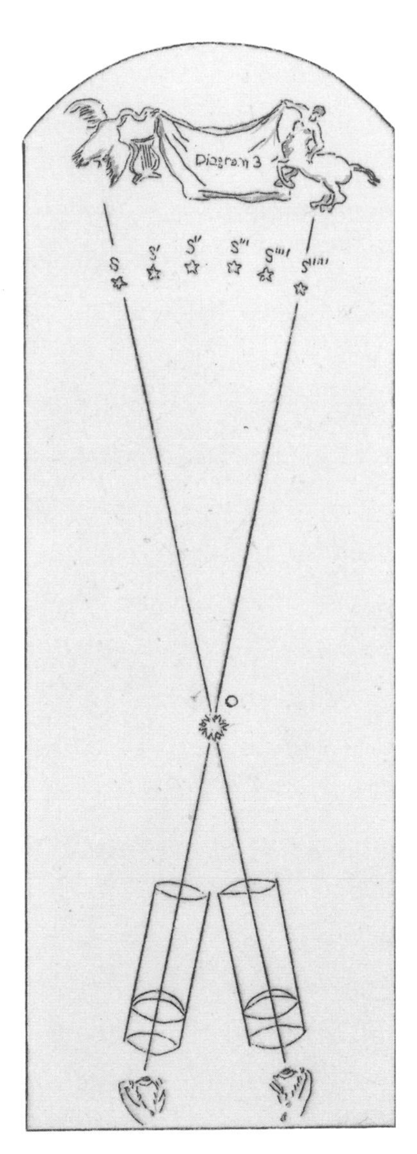

Diagram 3 shows the reasoning of Bessel, von Struve, and Henderson. The points A and B are the positions of the earth six months apart, for example in January and July. S, S′, S″, etc. are dim stars which can reasonably be assumed to be so remote that their positions look the same whether viewed from A or from B. O is a bright star, suspected to be much nearer the earth than S, S′, S′′, etc. Furthermore, O is chosen to be a star situated far from the ecliptic. The reason for this choice is that the ecliptic is the apparent path of the sun among the stars. If O were chosen to lie on the ecliptic, the extension of a striaght line from A to B might pass through O, and therefore O would not appear to shift when the earth moves from A to B. As Diagram 3 shows, the fact that O does not lie on the same line as A and B ensures a shift of its position relative to the background S, S′, S′′, etc., as the earth moves from A to B. The crucial question is whether the shift is large enough to be seen and measured. By 1838 instruments and techniques of observation had improved so much that Bessel, von Struve, and Henderson did observe shifts. In all three cases the shifts were less than one second of arc. A second is one-sixtieth of a minute and a minute is one-sixtieth of a degree (by contrast, Tycho could only measure shifts of a few minutes of arc or more). The observed shifts permitted the astronomers to calculate the distrances of the three stars, the closest turning out to be Alpha Centauri, whose distance is about 135,000 times the diameter of the orbit of the earth about the sun. Their observations and calculations confirmed the answer that Master Vittorio, following Copernicus and Aristarchus, had given to Tycho: that the stars are much farther from the earth than the outermost planet. And the shift of the positions of the bright stars relative to the background S, S′, S′′, etc provided excellent evidence for the motion of the earth.

Another objection to Copernicus's theory was based on the well known centrifugal effect: that a stone attached to a cord and whirled in a circle exerts a force on the cord and can even break it. Critics of Copernicus argued that if the earth rotated, then bodies attached to the earth would exert forces on the attachments, and if the attachments were broken the bodies would be hurled off the earth's surface. Since unattached objects are not in fact hurled off the surace of the earth, the critics concluded that the earth is not rotating. Master Vittorio dismissed this objection by pointing out that even without cords bodies on the surface of the earth are attracted towards the center of the earth by gravity, and even though he did not have a quantitative theory of gravity he was confident that gravitational attraction was sufficient to compensate for the centrifugal force on bodies due to the earth's rotation. The theory of gravitation that Vittorio hoped for was worked out by Isaac Newton in his *Mathematical Principles of Natural Philosophy,* published in 1687. Newton proposed in that magnificent book the law that any two bodies attract each other by a force proportional to the product of their masses and inversely proportional to the square of their distances. From this law of gravitation, together with general principles of mechanics, he was able to explain the motions of the moon, planets, comets,

seas, and bodies falling near the earth's surface. More will be said later about Newton's analysis of the motions of the planets. What we want to stress now is the ratio of the gravitational attraction the earth exerts upon a body near its surface to the centrifugal force on the body when it rotates along with the earth. The centrifugal force is greatest at the equator because the circle around which it moves is larger there than at higher latitudes. But even at the equator, the measured weight of the body is about three hundred times the centrifugal force. For this reason, a body lying on the surface of the earth without cords, attracted only by gravitation, does not fly off centrifugally. This does not imply that the centrifugal force has no consequences. At the north pole the centrifugal force is zero because the circle in which the body is whirled by the rotation of the earth is shruken to a point. Therefore, a body stretches the spring of a scale more when it is weighed at the north pole than when it is weighed at the equator. Part of the difference in weight is because the earth is flattened at the poles so that the body is closer to the center of gravity of the earth when it is at the poles than when it is at the equator. Most of the difference in weight, however, is due to the presence of a centrifugal force in the opposite direction from the earth's gravitation pull when the body is at the equator, and the absence of the centrifugal force when the body is at the pole.

It should be added that the flattening of the earth at the poles can also be understood as the consequence of the rotation of the earth. Long after Newton's time astronomers and geologists collected evidence that billions of years ago the matter of the earth was flattened at the poles. Newton had already reasoned that a rotating ball of fluid would bulge at the equator. Furthermore, even before precise measurements were made of the shape of the earth, he had an argument that the earth is flattened at the poles: that if the earth were a perfect sphere or were elongated at the poles then the seas would drain away from the polar regions and rise up near the equator covering all the lands there with water, which we know is not true. It was not easy to perform the measurements that establish the shape of the earth. The best measurements in the early eighteenth century were by Jacques Cassini, and they seemed to show that the earth is elongated at the poles. This result was disturbing because of the theoretical reasons for the opposite conclusion. This discrepancy between theory and observation motivated the French government to send two expeditions armed with excellent scientific personnel to measure the length along the surface of the earth of a degree of latitude both near the equator and in the arctic region. The latter turned out to be the longer, as the hypothesis of flattening at the poles requires. When the arctic expedition returned to Paris in 1737, its leader, Pierre de Maupertuis, was hailed as "the flattener of the earth and of Cassini."

In 1851 Leon Foucault exhibited some new evidence for the rotation of the earth that would have delighted Master Vittorio. He constructed a frame more than sixty meters in height from which a wire holding a heavy metal ball was suspended. The ball was made to swing like a pendulum in an initial vertical plane. At first the ball seemed to continue swinging in this initial plane, but gradually the plane of swinging was seen to rotate. This phenomenon would be completely mysterious if the earth were stationary, but it can be understood in the light of

the rotation of the earth. In Diagram 4 Foucault's apparatus is taken for simplicity to be installed at the north pole. If the pendulum is assumed to remain swinging in its initial plane relative to the stars, then the rotation of the earth beneath the swinging pendulum will make the plane of the swinging pendulum rotate at a uniform rate relative to the surface of the earth. Since Foucault actually installed his apparatus in Paris, far from the poles, this simple description of the motion of the pendulum has to be modified, but it still was the case that the plane of the swinging pendulum rotated relative to the earth's surface. Foucault pendulums can be seen today in various science museums, and their behavior continues to seem strange even when the phenomenon is understood.

One flaw in Copernicus's work troubled Master Vittorio more and more as time went by and new observational data, especially from Tycho Brahe's observatory, became known. Copernicus had assumed that the motions of the planets about the sun either were uniform circular motions (unifom in the sense that equal arcs of the circle are traversed in equal intervals of time) or else were compounded out of uniform circular motions. Displacement of the sun from the center of a circular orbit was permitted in his scheme and gave him some additional flexibility in making a geometrical model of the apparent positions of the celestial bodies in the heavens. The apparent positions of some of the planets, especially Venus, could be fitted very well by means of combinations of uniform circular motions, but this was not the case with Mars and Mercury. Much of Master Vittorio's time when he was not teaching was devoted to dreaming up more and more elaborate arrangements of what he called "wheels," in order to justify Copernicus's theory of circular motions. After years of futile calculations,

Master Vittorio learned that Johannes Kepler, who had been an assistant to Tycho Brahe in Prague and after Tycho's death in 1601 had succeeded him as mathematician to the Emperor Rudolph II, had also been analyzing the data concerning Mars and had achieved a remarkable success. For many years Kepler tried to to fit the observed positions of Mars with variants of the theory that earth and Mars moved in eccentric circles about the sun—that is, with the sun not at the center of the circles. His best model succeeded in fitting the positions of Mars observed by Tycho Brahe with an accuracy of eight minutes or better. By the standards of the theories of Kepler's predecessors this fit of theory to data was astonishing and would have satisfied any astronomer with a less rigid intellectual conscience. But Kepler deeply respected the data of Tycho Brahe, which he had reason to believe were accurate to within three or four minutes, and he also imposed very high standards on himself. In desperation he began to try to fit the data by supposing an oval orbit of the planet and eventually was led to a theory that fit the data with great accuracy. According to his theory, Mars moved in an elliptical orbit with the sun at one focus of the ellipse, as depicted in Diagram 5. It did not move uniformly in length along the elliptical path nor did the radius from the sun to the planet move through equal angles in equal intervals of time. Instead, as shown in Diagram 5, the motion is uniform in another way: equal areas are swept out by the line from the sun to Mars in equal intervals of time. In order to plot

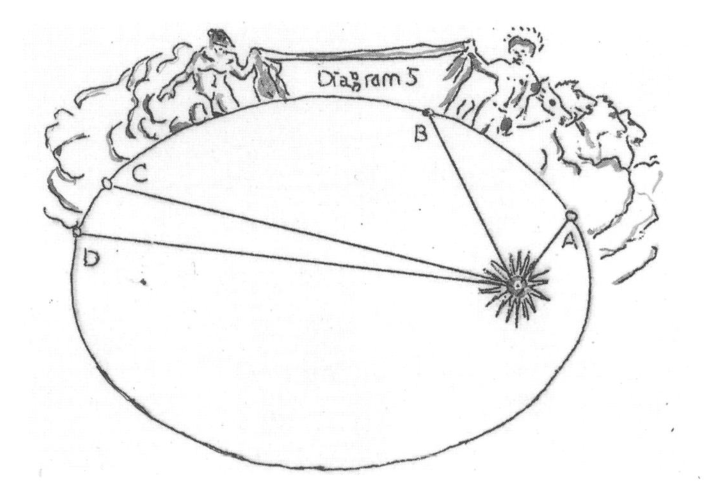

the apparent positions of Mars relative to the earth, Kepler also proposed that the orbit of the earth is elliptical with the sun at one focus, and the same law concerning equal areas swept out in equal times held for the earth as for Mars. But, the orbit of the earth was less elongated, hence closer to being a circle, than the orbit of Mars. Kepler's accomplishment was revolutionary in two ways. First of all, he freed astronomy from the doctrine, which had lasted at least two millennia, that the circle is the natural figure of the motion of heavenly bodies. And secondly, he gave unprecedented respect to the observational data. Kepler's accomplishment appears even more awesome when one considers the vast quantity of detailed and difficult calculations that he had to perform by hand.

No one was more impressed by Kepler's work than Master Vittorio when he read the book on the motion of Mars that Kepler published in 1609. He was so exhilarated that he traveled to Prague in order to offer his services, and just as his grandfather had been Copernicus's assistant, he became Kepler's. He assisted Kepler in the calculations showing that all other known planets—Mercury, Venus, Jupiter, and Saturn—obeyed the law of motion in an ellipse with the sun at one focus and the law of equal areas swept out in equal intervals of time. He also participated in the calculations establishing Kepler's third law of planetary motion, which related the period of a planet's orbit to its distance from the sun.

After leaving Bologna, Master Vittorio corresponded occasionally with Tibaldo since they had remained friends after Tibaldo's graduation from the school of St.-Joseph-in-the-Corner. Most of the correspondence was about such matters as planets, telescopes, and optics. But Vittorio also told Tibaldo that Kepler had tried to persuade the Protestant states of the Holy Roman Empire to adopt the Gregorian calendar, because of its obvious superiority to the Julian calendar but was unsuccessful on account of intense feelings against the Pope. One of Vittorio's letters remarked that Kepler was very skeptical of astrology but nevertheless obliged to make astrological predictions for Emperor Rudolph II and court officials. He quoted a statement of Kepler that, "Nature, which has conferred upon every animal the means of subsistence, has given astrology as an adjunct and ally to astronomy." Vittorio wrote, "As for me, I never would stoop to astrology. Before I would draw up ridiculous horoscopes for superstitious noblemen I would make my living as a shepherd." Tibaldo wrote back to him,

"Master, it is better that you not become a shepherd. Your sheep would all stray away while you scan the heavens for comets."

One more thing troubled Master Vittorio to the end of his life, something that he referred to as "the Osiander scandal." Copernicus finished his manuscript in his last year and was to ill to supervise its publication. He entrusted it to Vittorio's grandfather, Georg Joachim Rhaeticus, who in turn needed some help from Andreas Osiander, a Lutheran minister with an amateur interest in astronomy. Osiander feared that Copernicus would be attacked for his revolutionary proposition that the earth revolves about the sun and therefore wrote a preface stating that the proposition was nothing more than a "hypothesis" used to simplify astronomical calculations. Since the preface was anonymous an unwary reader could easily have inferred that it had been written by Copernicus himself. Master Vittorio was outraged at Osiander's sly attempt to misrepresent Copernicus's deeply held beliefs. But what disturbed Master Vittorio even more than this personal affront to a man whom he revered was the reflection, "How can I show that Osiander is wrong? After all, any motions that can be described with the sun as the stationary point can equally well be described with the earth as the stationary point. There are infinitely many ways to represent a set of astronomical observations, just as there are infinitely many different places where the observer can stand? What makes one representation truer than another?"

No satisfactory answer was given to these questions until Newton's book appeared in 1687. Newton realized that any set of bodies that are at rest relative to each other can be used as a reference frame for describing any motions in nature. For the purpose of description one reference frame is as good as any other. So far, Osiander is justified. But different reference frames are definitely not equally good for studying the causes and effects of motions. For some reference frames Newton's laws of motion hold, in particular: (1) a body not acted upon by any force remains either at rest or in uniform motion in a straight line; and (2) the acceleration of a body is proportional to the force upon it. The reference frames for which Newton's laws do hold came, in later times, to be called *inertial frames*. It follows immediately that each inertial frame is at rest or in uniform motion in a straight line relative to each other inertial frame. Consequently, a reference frame that is accelerated—for instance, rotating relative to an inertial frame—cannot itself be an inertial frame. Once the distinction between inertial and noninertial reference frames has been made we can go back to Osiander's preface and ask, "Is a reference frame in which the earth is at rest an inertial frame? Is a reference in which the sun is at rest an inertial frame?" One might expect that Newton, who followed the great tradition of Copernicus and Kepler, would surely have answered no to the first question and yes to the second. But Newton surprises us, as he surprised the astronomers and physicists of his time. His answer to both question is no! What Newton's theory tells us is that a reference frame which is not rotating relative to the distant stars and in which the center of mass of the solar system is at rest is to a very good approximation an inertial frame. The earth obviously is accelerated relative to this frame because of its rotation about its axis and its revolution about the sun. But the sun also is accelerated relative to it, because the sun is accelerated by the gravitational attraction of the planets just as they are accelerated by the gravitational attraction of the sun. But since the sun contains much more mass than all the planets combined, it never is far from the center of mass of the solar system and its acceleration relative to the center of mass is much less than the typical accelerations of the planets. For this reason a reference frame which does not rotate relative to the fdistant stars and in which the center of the sun is at rest is much closer to being an inertial frame than a frame of reference in which the earth is at rest. In this way and with these reservations Newton furnished the answer to Osiander that Master Vittorio had tried to find. But providing this answer was only a small part of Newton's accomplishment. From the laws of motion and the law of gravitation, together with the fact that the mass of the sun is much greater than that of all

the planets combined, Newton was able to make mathematical derivations of the motions of the planets. He showed not only that to a very good approximation the planets revolved in the elliptical orbits, which Kepler had inferred from analyzing astronomical data, but that certain corrections of Kepler fitted later and more accurate data. For example, he wrote to the Astronomer Royal asking if Saturn's motion deviated from the Keplerian orbit when it was close to the massive planet Jupiter, and the Astronomer Royal found the predicted deviation when he looked carefully for it.

Messages from the Stars

Master Vittorio lived in a great age of astronomy. In 1609, the year that Kepler published his discovery of the elliptical orbits of Mars and the earth, Galileo Galilei heard about the invention in Holland of a simple telescope, which made distant objects seem closer. He understood that such an instrument would be extremely valuable for astronomical observation and experimented with lenses fixed in tubes until he succeeded in constructing a telescope that made a distant object appear to be one-thirtieth of its true distance from the observer. Several other scientists independently constructed telescopes for astronomical purposes at nearly the same time, but Galileo made the most systematic and surprising use of the new instrument. In 1610 he announced his discoveries in a small book called *Sidereus Nuncius* (usually translated as *The Starry Messenger,* but the Latin word *nuncius* means either *message* or *messenger*). Galileo observed innumerable stars that were invisible to the naked eye. His telescope revealed that the luminous band across the sky known as the Milky Way actually consists of distinct stars which are so numerous that they create the appearance of an immense continuous source of light. He showed that the moon is not at all a smooth perfect sphere but contains mountains and valleys whose height and depth he could estimate by observing shadows. He found that the planet Jupiter has four large satellites which follow Jupiter as it moves against the background of stars while revolving about Jupiter in different orbits. Galileo would have been astonished and delighted to learn that in 1996 a spacecraft named *Galileo* was launched into space from the earth, traveled to the neighborhood of Jupiter, and observed at close range the four satellites that he was the first to notice and the smaller satellites that were discovered since his time.

Galileo also published letters reporting discoveries made with his telescope of spots on the surface of the sun. He did not pretend to know exactly what the spots are but he did observe that they changed their shapes, sizes, and brightness from day to day. He also found that despite irregularities of the motions of the spots, there was enough agreement among the motions of different spots to indicate that the sun as a whole is a rotating body. These observations were shocking to scientists and philosophers who had inherited from antiquity the idea that the matter of the celestial bodies differs from earthly matter by being perfectly pure, unchanging, and incorruptible. Galileo's conservative opponents tried to show that his observations were illusions or that his conclusions about the imperfections of the sun's surface were erroneous, but eventually his discovery was generally recognized by astronomers. However, personal troubles accompanied Galileo's scientific triumphs. He suffered from condemnation by the Church for his defense of Copernican astronomy, and from loss of eyesight in his old age probably due to looking at the sun through his telescope.

Astronomy has received innumerable messages from the stars in the period between Galileo and the present, some of them even more astonishing than the ones presented in the *Sidereus Nuncius*. After evidence had been presented that celestial matter shares with earthly matter the property of changeability, thoughtful scientists were inevitably curious about the exact nature of the matter of the stars. But because of the immense distance between the earth and the stars the answer to this question seemed beyond the bounds of human knowledge. By the middle of the nineteeth century, however, it was found that starlight carries messages about the material composition of the stars, and much other information besides.

The first step to an answer was an observation by William Wollaston in 1802 that sunlight refracted into a spread of colored light by a prism exhibited the oddity of some narrow dark lines. This phenomenon was investigated more systematically in 1814 by Joseph Fraunhofer, who observed the refracted light by means of a telescope and mapped more than 750 dark lines in the spectrum of the sun. Furthermore, in 1823 Fraunhofer discovered dark lines in stellar spectra. Explanations were provided in 1859 by Gustav Kirchhoff who made two great discoveries: (1) each chemical element produces a characteristic array of colors of light when samples are vaporized and heated enough to emit a considerable amount of light; and (2) when light having a continuous array of colors passes through a vapor of some chemical element it absorbs exactly those colors which it would emit if it were heated and used as a source of light. Hence, by absorption the vapor of a chemical element leaves characteristic "finger prints" on light passing through it. Most of the dark lines observed by Wollaston and Fraunhofer turned out to correspond to the characteristic colors of known chemical elements, such as hydrogen, oxygen, sodium, and calcium. It was later found by Norman Lockyer that there are some dark lines in the solar spectrum that did not correspond to the colors emitted by any known chemical element. He guessed that these were due to a hitherto unknown chemical element, which he called helium, (named for the Sun—*helios* in Greek). It did not turn out, however, that helium is a kind of matter found only in celestial bodies and not on earth because at the end of the nineteenth century laboratory samples of helium were discovered in emissions from radioactive elements.

A more complete understanding of the dark lines of the solar and stellar spectra required attention to the temperatures of the objects. The celestial objects that shine by their own light are intensely hot in their inner layers. It has been calculated, for example, that in order for the pressure inside the sun to be high enough to prevent the gravitational collapse of such an enormously massive body the interior temperature must be of the order of ten million degrees centigrade. At this temperature the velocities of the atoms are so great that their electrons are stripped off by atomic collisions and most of the light produced is due to collisions of the free electrons with the ionized atoms. This mode of generation of light produces a continuum of colors. By contrast, in typical laboratory experiments with vaporized samples of a chemical element, almost all the atoms are un-ionized—that is,

equipped with their normal set of electrons—and these atoms produce the spectral lines characteristic of the chemical element. Diagram 6a shows light with mixed

colors passing through a prism refracting different colors by different amounts and spreading them out on a viewing screen.

Diagram 6b shows a small sample of the display on the screen: two sharp yellow lines (the so-called D lines) which are among those emitted by hot un-ionized sodium vapor.

Diagram 6c shows light with a continuum of colors produced on the left by a body of extremely hot ionizied sodium atoms; when this light passes through the cool sodium vapor on the right the D lines (as well as other

colors which un-ionized solium atoms are capable of emitting) are absorbed, leaving dark lines in the continuum of colors coming from the left. Surrounding the hot interior of the sun is a layer of mainly un-ionized atoms with a temperature of 5000 to 10,000 degrees centigrade, which in accordance with Kirchhoff's law absorb the colors that they are capable of emitting. The same explanation applies to luminous stars other than the sun, though of course there are great differences from one star to another regarding temperatures, chemical elements that are present, and their abundances. In this way the light from a star carries a message about its material composition, and a series of dedicated physicists, chemists, and astronomers learned how to read the message. The light from a star can also contain a message about the speed of the star towards or away from the earth. The message is due to the fact that light, like all kinds of wave motion, exhibits the *Doppler effect*. This effect is the increase of frequency of the wave motion if the distance between the source and the observer is decreasing, and the decrease in frequency if the distance between the source and the observer is increasing. The reason for the change of frequency is that more complete periods of the wave motion reach the observer per second when the relative distance is diminishing than when the distance is constant, and fewer complete periods reach the observer per second when the distance is increasing. Many people who have listened to locomotive whistles are familiar with this effect, because the pitch of the whistle increases with frequency and is therefore higher when the locomotive is approaching the observer than when it is departing. Light is also a wave phenomenon, with violet light having higher frequency than blue, blue having higher frequency than green, and so on down through yellow, orange, and red. Consequently, if a source of yellow light is approaching the observer the Doppler effect will make the color shift in the direction of violet, the exact amount of the shift depending upon the speed of approach, whereas if the source is departing from the observer its color is shifted towards the red end of the spectrum. The Doppler effect also occurs in light which is invisible to human eye, both in infrared light with frequencies lower than that of visible light and in ultraviolet light with frequencies higher than that of visible light.

A dramatic example of the Doppler effect is the case of double stars of roughly the same chemical compositions and temperatures, such as Mizar in the Big Dipper. Mizar really consists of two stars which revolve around each other once in about twenty and a half days, and each star by itself produces approximately the same spectrum. At the time when both stars are moving perpendicularly to their line of sight to the earth, neither is approaching or receding from the earth and hence an earth-bound observer sees no Doppler shift in the light that they emit. The frequencies of light from each star are the same and their spectral lines coincide, as shown in Diagram 7a. When, however, the two stars are moving along the line of sight to the earth, then one must be approaching and the other is receding, and the frequencies of light from the approaching star are increased and those of light from the receding star are decreased. The spectrum produced at this time in the cycle by Mizar as a whole consists of many pairs of lines, with the members of each pair separated in frequency from each other by the amount of the Doppler shift (Diagram 7b). When the spectrum from a star varies over a period of time between the pattern of Diagram 7a and the pattern of Diagram 7b one has a clear message that the star is not a single but a double star, even when the distance between them is so small that one cannot see them separately with a telescope. When the two members of a double star are not as similar as those in Mizar, for example when one is luminous and one is dark, a message is still sent by the Doppler shift because the spectral lines of the luminous star vary in frequency during the period of revolution.

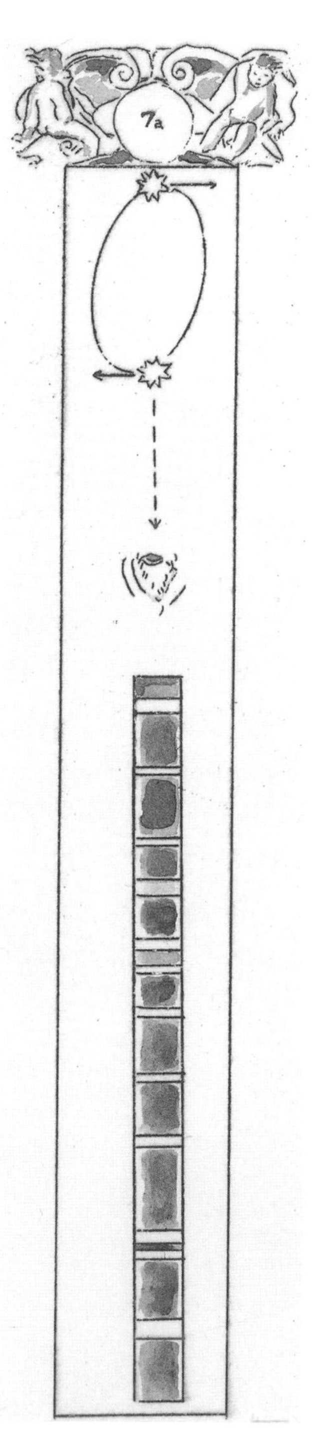

In the 1920's Edwin Hubble found that the more distant a

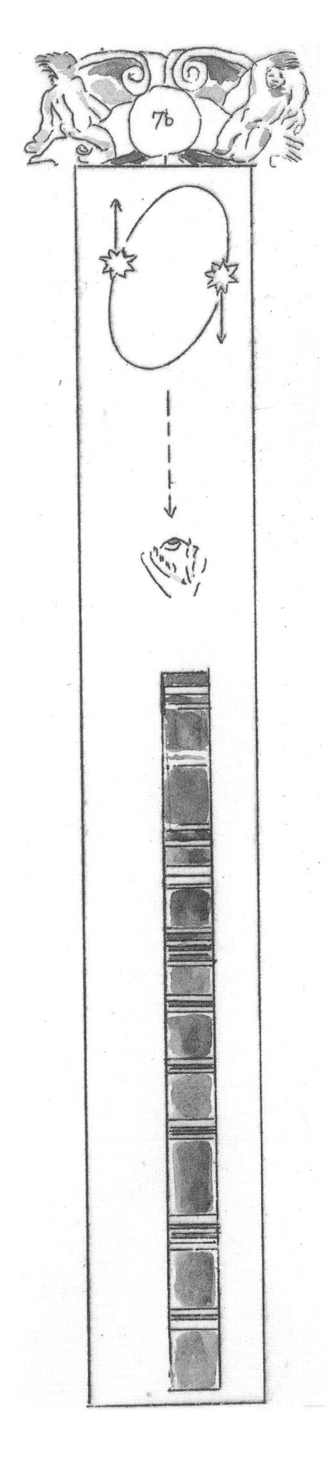

galaxy is from the earth the greater is the Doppler shift toward the red end of the spectrum. In other words, the more distant a galaxy is from us, the faster it is receding from us. If one combines this observation with the proposition that the earth has no privileged place in the universe—which after all was part of the lesson of Copernicus—then an observer on each galaxy must see the distant galaxies receding. The simplest explanation, accepted by almost all astronomers, is that the universe as a whole is expanding. But then there must have been a time when the matter of the universe was compressed into an extremely small region compared with the size of the universe indicated by the most powerful telescopes. There is much disagreement among astronomers and physicsists about the characteristics of matter when the universe was immensely denser and hotter than at present, and even disagreements about whether the laws of physics were the same then as now. There is a general agreement, however, that an immense explosion occurred about 15 to 20 billion years ago, far exceeding even the supernovas which emit in a few hours as much energy as the sun radiates in a hundred million years. There is also agreement that such an explosion, usually called *The Big Bang,* would leave a record in the form of radiation detectable in each spatial direction regardless of the presence or absence of a galaxy in that direction. Such radiation was discovered in 1965 by Arno Penzias and Robert Wilson using not an optical telescope, but a radio telescope. It was necessary for them to employ this new kind of instrument because the radiation left over from the Big Bang has frequencies far below those of visible light. The message they received came from every direction in space, but most important, it was a message from the beginning of the physical universe.

Master Vittorio was right when he told his class, *If we could only live four hundred years, what glorious discoveries we would witness!* But there are so many things that we still do not know about the remote galaxies and the early universe that he could be saying the same thing today.